Brief Introduction to Fisheries

Xinjun Chen • Yingqi Zhou

Editors

Brief Introduction to Fisheries

Editors
Xinjun Chen
College of Marine Sciences
Shanghai Ocean University
Lingang New City, Shanghai, China

Yingqi Zhou
College of Marine Sciences
Shanghai Ocean University
Lingang New City, Shanghai, China

ISBN 978-981-15-3334-1   ISBN 978-981-15-3336-5   (eBook)
https://doi.org/10.1007/978-981-15-3336-5

This Springer imprint is published by the registered company Springer Nature Singapore Pte Ltd.
The registered company address is: 152 Beach Road, #21-01/04 Gateway East, Singapore 189721, Singapore

Contents

1   **Introduction** 1
Xinjun Chen and Yinqi Zhou

2   **Review on Global Fisheries** 19
Xinjun Chen, Yinqi Zhou, and Leilei Zou

3   **Overview of Major Fishery Countries in the World** 97
Jiahua Le and Xinjun Chen

4   **An Overview of Major Fishery Disciplines** .131
Xinjun Chen, Yinqi Zhou, Liming Song, Zhihe Wang, Kaijun Wu,
and Chen Sun

5   **Sustainable Development and Blue Growth of Fisheries** 181
Xinjun Chen and Yinqi Zhou

6   **Global Environmental Change and Fisheries** 225
Xinjun Chen

Chapter 1
Introduction

Xinjun Chen and Yinqi Zhou

Abbreviations

COFI   The Committee on Fisheries
FAO   The Food and Agriculture Organization of the United Nations

1.1 Definition and Features of Fisheries

1.1.1 Definition and Classification of Fisheries

1.1.1.1 Definition of Fisheries

Fishery is an important industry in the world and plays an important role in the human welfare. In Asian countries and regions such as China, Japan, and South Korea, it is customary to refer to fisheries as the aquaculture and fishing industry (Chen and Zhou 2018). According to the definition given by the *Chinese Agricultural Encyclopedia*, "fishing industry" refers to "the social industrial sector where human beings obtain aquatic products through fishing, aquaculture and enhancement that are produced under the material transformation function of biological mechanisms in the water" (Chen and Zhou 2018). In China, fisheries, in a broad sense, also include the repair and construction of fishing vessels, design and manufacture of fishing gear, construction and planning of fishing ports, and supply of fishery products; in addition, fisheries involve various pre- and post-harvest industries, which include processing, preservation, storage, transportation and sales,

X. Chen (✉) · Y. Zhou
College of Marine Sciences, Shanghai Ocean University, Lingang New City, Shanghai, China
e-mail: xjchen@shou.edu.cn; yqzhou@shou.edu.cn

© Science Press & Springer Nature Singapore Pte Ltd. 2020
X. Chen, Y. Zhou (eds.), *Brief Introduction to Fisheries*,
https://doi.org/10.1007/978-981-15-3336-5_1

cultivation, and harvest of aquatic products. In China, fishery is an important industry that is listed as one of the components of agriculture (Chen and Zhou 2018). However, in Western countries and regions, such as Europe, it is customary to refer to fisheries as the fishing industry and aquatic products processing industry, and the industrial chain of fishing, processing, storage, and transportation and sales is viewed as a complete industry. Fisheries mean the combination of industries from the development and utilization of natural resources to fishing for aquatic living resources where the end consumers are viewed as service targets and are thus called "the fishing industry." At the same time, aquaculture is treated as a sideline of agriculture and has not been specifically classified as an industry (Chen and Zhou 2018). For a long time, the Committee on Fisheries (COFI), established by the Food and Agriculture Organization of the United Nations (FAO), has focused on and coordinated national fishing-related activities. Therefore, as far as "marine fishery" is concerned, this term customarily refers to marine fishing and production and the related aquatic products processing industry, while marine aquaculture is not recorded in a marine fishery. Until the end of the twentieth century, when the global aquaculture industry rapidly developed, the output and output values continued to rise and aquaculture products increasingly contributed to the protein and economies of human society; COFI established the Committee on Aquaculture in 2000 (Committee on Aquaculture-COFI/FAO). Over the past 10 years, inshore marine cage aquaculture has rapidly developed, its production has soared, the resulting proportion in marine fisheries has increased, and culture species, including traditional fishing targets, have attracted widespread attention. Therefore, the international community is accustomed to referring to fishing and aquaculture as "fisheries and aquaculture." However, in this text, we refer to fisheries as the aquaculture and fishing industry (Chen and Zhou 2018).

1.1.1.2 Classification of Fisheries

Fisheries are a production industry, involving the exploitation, rational harvesting, artificial enhancement, and aquaculture of aquatic products and animals that inhabit and breed in marine and inland waters and preservation and processing of aquatic products. In a broad sense, this industry also includes the repair and construction of fishing vessels, manufacture of related facilities and equipment, and production of fishery medicine and fish feed. Fisheries are a component of national economies. With the expansion of fisheries resources development in marine and inland waters and the increase in population, aquatic products have become not only an important source of animal protein but also the raw materials for chemical and pharmaceutical industries, and aquatic products have provided feed for the livestock industry as well.

Different nations have different fishery classifications. Taking China as an example, fisheries can be divided into the fishing industry, aquaculture and enhancement industry, and aquatic products preservation and processing industry. (1) Fishing industry: a production industry of fishing for naturally grown economic animals in marine and inland waters, including the marine fishing industry and inland water fishing industry. (2) Aquaculture and enhancement industry: a production industry for artificial breeding, culturing, releasing, and enhancing aquatic economic animals and plants in suitable inland waters, shallow seas, and mudflats. The former, through artificial culturing, is called aquaculture, and the latter, through natural breeding, is called enhancement. Fisheries can be divided by water type into inland aquaculture and enhancement and marine aquaculture and enhancement. The former cultures and enhances fish, shrimp, crabs, and turtles in ponds, lakes, reservoirs, rice fields, rivers, and other waters. The latter cultures and enhances shellfish, fish, shrimp, and crabs and cultures seaweed in shallow seas, mudflats, harbors, and other waters. (3) Aquatic product preservation and processing: a production industry for aquatic food preservation and processing and utilization of processed aquatic products. The former refers to freezing, refrigerating, marinating, drying, smoking, and canning aquatic products as well as their preservation and processing and production of various raw and cooked food in small packages. The latter refers to the production of pharmaceutical and chemical products, such as feed fish meal, fish oil, and cod liver oil, and the preparation of polyunsaturated fatty acids, algin, and iodine. Aquatic products preservation and processing plays a key role in promoting the circulation of fishing products and aquaculture products and in improving the edible value and utilization of aquatic products.

In addition, there are cultivated fisheries, recreational fisheries, and urban fisheries as well. (1) Cultivated fishery: also known as the "marine enhancement industry". This marine fishery uses a similar production model as agriculture and animal husbandry in suitable waters and is a new system of marine biological resource development, utilization, and management integrated with marine fishing and marine culturing. The cultivated fishery is a system that cultures seaweed and cultures and enhances fish, crabs, shrimp, and shellfish by applying modern science and technology and equipment and adopting technical measures such as artificial incubation, breeding, release, and artificial reefs. The cultivated fishery attaches great importance to protecting the propagation of aquatic resources, promoting aquatic productivity, and maintaining ecological equilibrium. (2) Recreational fishery: based on leisure industries such as tourism, fishing, entertainment, catering, fitness, and vacation. The recreational fishery forms a new industry integrated with components of tourism, recreation, and fishing. The recreational fishery has achieved the purpose of interacting with primary, secondary, and tertiary industries to improve the social, ecological, and economic benefits of fisheries to meet the increasingly spiritual and cultural needs of human beings. In the USA, Japan, and European countries, recreational fisheries are well developed. As China has been rapidly developing the recreational fishing industry since the twenty-first century, it takes its place in fisheries. (3) Urban fishery: takes advantage of the economy, culture, science, and technology in large cities to develop a production industry of intensive fishing to meet the consumer needs in large cities. The urban fishery upgrades and expands traditional fishing and is a component of urban agriculture, a modern fishing pattern with suburban characteristics, and special urban service functions.

Fisheries can be divided into marine fisheries and inland fisheries by different water types. (1) Marine fisheries can be divided into coastal fisheries, inshore fisheries, offshore fisheries, and deep-sea fisheries. Deep-sea fisheries can be further divided into distant water fisheries and high seas fisheries, and high seas fisheries are also known as oceanic fisheries. Meanwhile, marine fisheries can be divided into aquaculture and enhancement as well as capture, as the former can be divided into the marine resource enhancement industry and marine aquaculture industry. The latter refers to a production industry that focuses on sustainable development and rational utilization of marine fishery resources. Marine fisheries can be divided into coastal fishing, inshore fishing, and deep-sea fishing according to different waters. Modern marine fisheries, in a broad sense, also include fishery products preservation, processing, distribution, trade, and so on. (2) Inland fisheries are conducted in inland ponds, lakes, reservoirs, rivers, paddy fields, and so on. Inland fisheries can be further divided into freshwater fisheries and reservoir fisheries and can also be divided into inland aquaculture and enhancement as well as inland capture. Not all inland waters are freshwater, and there are still many saltwater lakes; even so, inland fisheries are often known as freshwater fisheries.

Additionally, marine fisheries can be divided into commercial fisheries and small-scale fisheries. In addition to commercial fisheries, there are also subsistence fisheries, where fish and shrimp are captured mainly for household consumption and only a small amount are for sale in exchange for necessities, so subsistence fisheries involve fishermen exchanging goods for goods to make a living. The international community has given special attention and protection to the fishery rights and interests of subsistence fisheries fishermen.

Traditionally, fisheries are classified and named according to the various aquatic species, operating methods, or waters, such as squid fisheries, tuna fisheries, trawling fisheries, purse seine fisheries, and fixed net fisheries.

1.1.2   Characteristics of Fisheries

1.1.2.1   Characteristics Originating from Natural Resource and Environmental Features

Seasonality

Fisheries production targets biological resources in water, so it is obviously seasonal (Chen 2014). The most influential factors on fishermen are longer production cycles and concentrated and short-term harvests or fishing seasons. Aquatic products are seasonal and perishable; thus, a greater capability is required for the concentration of aquatic products processing and preservation to sell products in a timely and regular faction. However, there is a contradiction between the supply-demand and concentrated fish harvest, which leads to a huge production capacity, low production equipment efficiency, and catch waste. The seasonal characteristic of seafood hinders aquatic products processing and preservation, industry combination, and function; thus, determining how to optimize organization to improve overall effectiveness and efficiency is a challenge facing the aquatic products industry chain.

Regionality

Regional distribution is a common characteristic of living species. Different species inhabit different waters and different water layers; even the same species can have different qualities and flavors due to various environments, which form region-specific specialties (Chen 2014). Aquatic products are obviously regional, and compared with other agricultural products, consumers have diverse demands for aquatic products and are particularly concerned about the origin and variety of aquatic products. Origin is often closely related to an aquatic product brand, so that products have geographical characteristics, such as Yangcheng Lake hairy crabs. Therefore, the regional characteristic of aquatic products and consumer focus on the origin are aspects that should garner attention during industrial development and management of fisheries. In addition, from the perspective of resource protection and conservation management, to protect and strengthen the supervision of fishery resources in typical waters, international fisheries management organizations require that aquatic products need to be accompanied with origin certificates and biological labels.

Shared

Aquatic living resources such as fish live and migrate in waters, where they even migrate across oceans and borders, and this liquidity has made fishery resources public resources (Chen 2014). Because of this liquidity or migration across borders, it is difficult to define resource ownership; thus, it easily leads to predatory fishing. Therefore, in fisheries resource management, to achieve sustainable use of fishery resources and sustainable development of fisheries, it is required that those who use fishery resources should cooperate, including cooperation and coordination among countries. For example, tuna fisheries, saury fisheries, and so on, the fisheries under the jurisdiction of various regional international fisheries organizations, are the responsible fisheries that are currently advocated internationally. However, fishery resources are shared, and ownership is ambiguous; thus, the priority for fishermen is resource possession, which leads to competition for resources. The fisheries industry has a strong exclusivity. At the same time, a good knowledge of fishery resources and good control of fishery resources are the core competitive factors.

1.1.2.2 Characteristics of the Fisheries Industry

Fisheries Industry Closely Related to "Resources, Environment, and Food Security"

As we all know, "resources, environment, and food safety" are world hot spots, which attract leaders' attention in different countries; thus, many summits and international seminars are held to put forward the guiding principles of sustainable development in social development (Chen and Zhou 2018). Sustainable development, harmonious resources and environment development, and human food security are all key issues in fisheries development. Therefore, the international fishing community, including governments, scientists, and the fishing industry, has worked out a series of international fisheries management agreements through consultation, and fisheries management has been enhanced through strengthening the role of regional international fisheries organizations. In China, although fisheries do not enjoy a prestigious position, any fishing activity is under the scrutiny and supervision of all society because it is closely related to resources, environment, and food security.

Comprehensive Benefits

Benefits of fisheries should be comprehensively evaluated from the combined economic, ecological, and social perspectives. In addition to economic benefits pursued by all industries, fisheries must focus on ecological and social benefits as well. Fisheries target renewable biological resources, and if we pay attention to fisheries resource conservation, it is possible to achieve sustainable development. However, if we merely emphasize the economic benefits and catch quantity, fishery resource exhaustion will accelerate, ultimately leading to industrial collapse. Therefore, paying attention to ecological benefits is in line with biological characteristics being renewable, but there is no denying that the priority for ecological benefits is at the cost of some economic benefits.

In addition, fish and other aquatic organisms are food as well as important sources of high-quality protein for human beings. In some areas, aquatic products are also important food sources and human livelihoods. Fisheries also provide jobs, and fishing villages and harbors are often gathering places for economic activities. Therefore, fisheries development is related to people's livelihood, and the social benefits of fisheries development are key issues deserving our concern. The introduction to the fishing access system and the licensing of aquaculture farm use promote coordinated development in fishery areas and protect fishermen's exclusive fishing rights. From the viewpoint of industrial management, the fisheries industry chain is so long that, from direct harvesting of natural resources and artificial culturing to processing and utilization and preservation and distribution, all processes have direct impacts on the overall benefits. Therefore, we shall pay special attention to coordinating and optimizing the industrial chain among various sectors to improve comprehensive benefits.

Joint Development with Other Industries

Many broad industries and techniques are involved in fisheries development. For example, fishing and production are closely related to engineering, environment, meteorology, biology, shipbuilding, electronic equipment, communication and information, mechanized equipment, synthetic fiber materials, processing and utilization, refrigeration, and so on; thus, fisheries development is highly relevant to technical developments in these fields. Without the support of shipbuilding, mechanical engineering, and electronics, it would be impossible to develop deep-sea fisheries. As another example, ultrasonic detectors, ship facilities, hydraulic machinery, and artificial and synthetic fibers greatly improve the efficiency of fishing operations. Fisheries development is closely reliant on the development of other industries; thus, fisheries are characterized by joint development with other industries. For this characteristic, we need to actively apply new scientific and technological achievements to fisheries, which are an important driving force for fisheries development (Chen and Zhou).

Perishability of Aquatic Product

Aquatic products are characterized as perishable, and product quality is closely related to conservation methods and techniques. Compared with other products, fishery products have high requirements for fishing technology, preservation, and logistics management. Aquatic products are for food consumption, so their safety must be guaranteed. This safety means that we cannot use a traditional management approach to randomly check product quality; instead, it is required that all products be reliable and consistently meet quality standards. Therefore, a traceable production management system that requires quality supervision and management for all aspects of the industry chain has been introduced. Moreover, a complete record, also known as a fisheries file, has been introduced as well. In the whole industry chain from raw materials to catches to consumers, technical measures must be taken to keep products fresh, alive, or frozen. However, the biodiversity of fish, shrimp, and other aquatic organisms makes technical measures for processing and conservation more complicated, and more technical measures need to be introduced for varieties of aquatic products. For example, tuna is most favorable to consumers when eaten raw; once captured, the fish should be immediately processed on board and rapidly frozen to $-60\ ^{\circ}\text{C}$. However, some products need to be kept fresh. In addition, perishability also affects the sales method of aquatic products (Chen and Zhou 2018). For example, in aquatic product wholesale auctions, the reverse auction method of pricing from high to low is adopted to ensure fish are sold quickly and to avoid abortive auction.

Instability of Fishery Industry

Fisheries target fish and other aquatic organisms; thus the change and fluctuations in resource make the fisheries industry unstable. Environmental and climate change also lead to habitat change and fluctuations in resource status, which cause different production statuses (Chen and Zhou 2018). To date, fisheries are highly dependent on nature, and the industry has high instability and risk in terms of the scale of production, planning, and economic benefits. Meanwhile, investment in fisheries is high risk, and fishing enterprises should pay special attention to risk aversion and speculation reduction.

Consumer Demand for Fish Diversity and Consumption Habits

Residents in coastal countries and regions around the world habitually consume fresh aquatic products. Compared to other agricultural and livestock products, consumers are more concerned about the species, origin, and production seasons of aquatic products, which have strong region-specific characteristics. In addition, consumption habits change with the times; thus, major changes will affect new consumption trends. Human beings are anticipated to have a gradually higher demand for processed or semi-processed products, which will become standard for daily consumption.

1.1.2.3 Development Trends of Fisheries

Industry Structure Transformation in Fisheries

The current global fisheries industry is witnessing a historic industrial transformation. Over the past 30 years, global fisheries production has been mainly from the fishing industry; however, during the last decade of the twentieth century, the rapid development of aquaculture has transformed the fishing industrial structure from "hunting" to "farming." The emergence of large marine cage aquaculture projects and land-based aquaculture projects is a symbol of modern fisheries in the twenty-first century (Chen and Zhou 2018).

Realization of Projectization Management in Fisheries

The goal of modern fisheries is to achieve projectization supported by technology. Projectization is mainly embodied in the fishing process and product standardization, carrying out the principles of quality and top-ranking efficiency. Standardization is based on quantitative control of the fishing process, i.e., digitization. Meanwhile, projectization is also reflected in the product quality guarantee, food safety, and traceability (Chen and Zhou 2018).

Prioritizing Comprehensive Benefits in Fisheries

Prioritizing comprehensive benefits is mainly embodied in extending the fishery industrial chain to enhance the overall comprehensive benefits. For example, recreational fisheries development combines material production with culture, recreation, and community development. As another example, aquatic products are not only important food for human beings but also important industrial raw materials. The overall economic benefits of fisheries can be improved through comprehensive utilization and deep processing, especially through developing marine medicine, biofuel, and biomass. Moreover, deep-sea fisheries are an industry resource, not only providing high-quality protein and food security for communities but also improving employment and international trade and reflecting the maritime rights and interests of countries and regions as well. Therefore, fisheries bring great economic, ecological, and social benefits (Chen and Zhou 2018).

1.2 Content and Disciplinary System of Introduction to Fisheries

1.2.1 Aims of Introduction to Fisheries

Introduction to fisheries is a fundamental course for comprehensive knowledge and quality education, an introductory course for those engaged in fisheries or fisheries-related jobs. This course focuses on the concept of fisheries, industry composition, and characteristics; fishery resources, environment, and sustainable development theory; fisheries subject composition, research, and scientific and technological achievements; fisheries development in the world's major countries; and the development status, focus, and trends of global fisheries. Introduction to fisheries is a typical comprehensive course, covering biology, economics, management, oceanography, sociology, and information economics and more. This course is tailored for undergraduates majoring in aquaculture or related majors, such as marine science and technology, marine management, marine science, fisheries economics, and so on.

1.2.2 Research Content of Introduction to Fisheries

Introduction to fisheries is an introductory course introducing the status of fisheries in national economies from the perspective of macro-strategic development; industrial structure and fishery characteristics; fisheries science and the branches and relationships therein; heated issues of international fisheries development, such as carbon sink fisheries; relations between fisheries and climate; and so on. This text essentially provides the basic concepts and knowledge for those engaged in fisheries or related jobs to master the correct methods of observing and researching fisheries issues.

This course consists of five parts: (1) concept of fisheries. This section includes the definition, industrial structure, characteristics, and roles of fisheries. (2) Fisheries and sustainable development. This section includes sustainable development theory, blue growth and international actions, and carbon sink fisheries. (3) Fisheries and science and technology. This section includes the fishery science system; influence of science and technology on aquaculture and enhancement; fishing, processing, utilization, and development of aquatic products; application of information technology in fisheries, fisheries economics, and fisheries management; and sustainable development in fisheries. (4) Status of world fisheries. This section includes the major global fishery resources and fishing grounds, evolution of global fisheries production, global fisheries production structure, global marine fishing, global aquaculture, global aquatic products processing industry, global modern recreational fisheries, global aquatic products trade, status and trends of international fisheries management, and current major issues and trends of global fisheries. (5) Status of fisheries development in major countries and regions. This section includes the status and role of fisheries in national and global economies, natural fisheries environments, fisheries species, and status and trends of fisheries development in major countries and regions (Chen and Zhou 2018).

1.2.3   Fisheries Discipline System and its Content

The fisheries discipline is also known as the "discipline of aquatic products." This comprehensive applied discipline conducts research into the rules of sustainable development and utilization of aquatic resources. This field mainly focuses on the growth and reproduction, distribution, and quantitative changes of aquatic economic animals and plants, fishing and catching and aquaculture and enhancement, theory and technology of aquatic products preservation and processing, design and application of related production tools and facilities, production operation and management, natural conditions, and human factors that affect production. The fisheries discipline is characterized by agronomy, engineering, management, and economics law. Subdisciplines under fisheries can be divided into aquatic resources, aquaculture and enhancement, fishing, aquatic products preservation and processing technology, aquatic engineering, fisheries economics, and fisheries management (Chen and Zhou 2018).

1.2.3.1   Discipline of Aquatic Resources

The discipline of aquatic resources is also known as the "discipline of fishery resources." This area is one of the subdisciplines of fisheries science, as well as an applied discipline focusing on the biological characteristics of aquatic resources, temporal and spatial distribution of biomes and groups, movement and migration, quantitative changes in population, relationships among species, aquatic resources evaluation, and relationships with environmental factors. This discipline is closely connected to the development of other disciplines, such as aquatic biology, ichthyology, hydrology, meteorology, and mathematical statistics. This field can provide a theoretical basis for the sustainable development and utilization of aquatic resources, fishing condition forecasting, and fisheries management measures. This area can be further divided into two disciplines concerning aquatic resources biology and aquatic resource evaluation.

1.2.3.2 Discipline of Aquaculture and Enhancement

Aquaculture and enhancement is one of the subdisciplines of fisheries science. This applied discipline focuses on the principles and technology of aquaculture and enhancement in natural or artificial waters and on interactions with the aquatic environment. This field provides the basis for expanding aquaculture-enhanced species, as well as improving their quality, enhancement effect, and technology. This area can be further divided into two disciplines concerning aquatic enhancement and aquaculture. Aquaculture can, through different research directions, be divided into the genetic breeding of aquatic animals, nutrition and feed of aquatic economic animals, and algae cultivation.

1.2.3.3 Piscatology

Piscatology, also known as "Discipline of fishing," is one of the disciplines of fisheries science. This applied science discipline focuses on the improvement of fishing gear and technology, formation mechanism of fishing grounds, and fish migration rules, according to fishing target species, life habits, distribution, and migration. This area provides the basis for the sustainable development and utilization of aquatic resources and development of the fishing industry. This field can be further divided into disciplines concerning fishing gear, fish ethology, fishing methodology, and fisheries hydrography. Fishing gear science conducts research into fishing gear design, material properties, and assembly technology. Fish ethological science conducts research into fishing targets. Fishing methodology science conducts research into fishing methods. Fisheries hydrography conducts research into the formation mechanism of fishing grounds.

1.2.3.4 Discipline of Technology for Aquatic Products Preservation and Processing

Technology for aquatic products preservation and processing is a subdiscipline of fisheries science. This applied science focuses on the properties, refrigeration and live-keeping, preservation and processing, and comprehensive utilization of aquatic products. This area provides the basis for improving the use efficiency of aquatic products and food value to meet human needs. This field can be further divided into disciplines concerning raw materials, aquatic food chemistry, refrigeration technology, and comprehensive utilization technology of aquatic products. Raw materials science of aquatic products studies the characteristics of aquatic products raw materials, food chemistry science of aquatic products studies the chemical characteristics of aquatic organisms, and refrigeration and comprehensive utilization technology science studies the preservation, processing, and comprehensive utilization of aquatic products.

1.2.3.5 Discipline of Fisheries Economics

Fisheries economics is a subdiscipline of fisheries science. This area is a cross-discipline between fisheries and sector economics, as well as an applied science focusing on economic relations and economic activity rules concerning production, allocation, and exchange and consumption. This subdiscipline provides a basis for establishing the scientific and rational aquatic economic system, production structure, and decision-making for sustainable aquaculture development and achieving optimum inputs and outputs. This field can be further divided into disciplines concerning aquatic resource economics, fisheries technology economics, and fisheries institutional economics.

1.2.3.6 Discipline of Fisheries Engineering

Fisheries engineering is a subdiscipline of fisheries science. This area is a cross-discipline between fisheries and engineering, as well as an applied science focusing on the characteristics, principles, and planning and design of relevant fishing facilities, equipment, and test instruments. This field can be further divided into disciplines concerning fishing vessel engineering, fishing port engineering, fishery machinery engineering, aquaculture engineering, processing engineering, and marine bioengineering.

1.2.3.7 Discipline of Fisheries Remote Sensing

Fisheries remote sensing is a cross-discipline between marine remote sensing and fisheries; analyzing, evaluating, and judging fishery resources quantity; and distribution and fishing grounds by using data on the surface temperature, water color, chlorophyll, and sea surface height, which are obtained from marine remote sensing satellites. This area is one of the research methods for fishery resources and fisheries oceanography. Marine remote sensing can synchronously and instantly collect environmental parameters for large ocean areas, which reflects the distribution characteristics of a marine environment, such as frontal zones and eddies. Therefore, this cross-discipline can preliminarily analyze and determine the distribution of regional economic animals, such as fish, and improve the ability to scout for fish stocks and explore fishing grounds.

1.2.3.8 Discipline of Fisheries Resource Economics

Fisheries resource economics is one subdiscipline of fisheries science. By using the basic economic principles, fisheries resource economics studies the contradiction between the demand of human economic activities and the supply of fishery resources in addition to studying the current and future optimal allocation of fishery resources and exploring problematic laws. This subdiscipline focuses on issues concerning fishery resources and fisheries resource economics. This field mainly solves the following problems: how to allocate current and future fishery resources; how to allocate the benefits generated by resource allocation decision-making among all members of society; what are the problems for fishery resource allocation and what are the reasons for these problems; what are the proposals and policies to address the above issues; and how to evaluate the benefits, costs, and impacts of these proposals and policies.

1.2.3.9 Discipline of Fisheries Laws and Regulations

The concept of "laws and regulations" of fisheries refers to laws and regulations in the broad sense. To put it simply, fisheries laws and regulations refer to the sum of fisheries-related legal norms, i.e., the general designation of legal norms adopted to adjust fisheries activities and relationships. Fisheries activities are mainly carried out in waters, and they have strong mobility characteristics. As a result, fisheries activities in the sea and inland waters adjoining multinational land territories are regulated by the UN Convention on the Law of the Sea and other relevant international laws. Therefore, fisheries laws and regulations, in this connotation, involve two major components: domestic fisheries laws and regulations under national legal systems and international fishing laws and regulations.

1.3 Aims of Learning Introduction to Fisheries and the Role of Fisheries

1.3.1 Aims of Learning Introduction to Fisheries

Introduction to Fisheries, as a discipline, systematically introduces the fisheries industry, status of global fisheries, fisheries scientific connotations and discipline system, science and technology, and fisheries development. This is a basic course for fisheries professionals. The aims and significance of Introduction to Fisheries are listed as follows:

1. To fully understand the nature and characteristics of fishery resources, the fishery industry structure, and fisheries discipline system, which lay the foundation for future work in fishing-related work

2. To have basic knowledge of the status of global fisheries resources, i.e., the status of fishery resources as well as the history and trends of fisheries development in major countries, which provides the basic knowledge for fisheries sustainable development
3. To have basic knowledge of how science and technology promotes global fisheries development, especially the development of newly emerging disciplines, such as science and technology development in the fishing and aquaculture industry, to provide the basis for grasping global fisheries development patterns
4. To learn what fisheries problems are confronted by countries around the world and what measures are being adopted to deal with these problems, such as the impact of global climate change on fisheries and aquaculture

1.3.2 Role of Fisheries in National Economies and Social Development

With the uncertainties brought about by climate change, economy, and finance as well as the increasingly severe competition for natural resources, human beings are currently faced with one of the toughest global challenges, that is, how to feed more than 9 billion people by 2050. In response, the international community made a commitment in November, 2015, which was declared in the *2030 sustainable development agenda*, a policy document approved by members of the United Nations. The *2030 sustainable development agenda* sets a goal for the contributions of fisheries and aquaculture to food security and nutrition and a code of conduct for natural resource utilization to ensure sustainable development of the economy, society, and environment.

Thousands of years after land-based food production shifted from hunting/collecting activities to agricultural activities, aquatic food production has also shifted from wild fishing-dominant fisheries to aquaculture-dominant fisheries with increasing options for farmed fish species. The year 2014 was a milestone year because this was the first year in which the contribution of aquaculture to human consumption of aquatic products surpassed the wild aquatic products fishing industry. Aquaculture is urgent and critical to meet people's growing demand for edible aquatic products as targeted by the *2030 Sustainable Development Agenda* (FAO 2016).

From the perspective of industry classification, fisheries is a sector of the national economy, like agriculture, forestry, mining, industry, commerce, and transportation. According to the FAO, fisheries are an important source of food and protein for billions of people around the world and maintain the livelihood of more than one-tenth of the population. Therefore, fisheries have an important position and play a key role in national economic development, food security, social employment, foreign exchange earnings, and social stability (FAO 2016).

1.3.2.1 Fisheries as Rich Sources of Protein

The role of fisheries in a national economy mainly lies in the fact that fisheries provide food for people, especially animal protein. When the livestock industry is underdeveloped, fish is the main source of animal protein. Aquatic products are very valuable sources of protein and essential trace elements to ensure balanced nutrition for the human body and to maintain good health. According to the FAO, the growth rate of the global fish food supply has exceeded that of the population for 50 years, from 1961 to 2013, with an average annual growth rate of 3.2% between 1961 and 2013, which was twice the population growth rate, thereby increasing the per capita consumption. Global per capita aquatic products consumption increased from 9.9 kg in the 1960s to 14.4 kg in the 1990s, then to 19.7 kg in 2013, and to more than 20 kg in 2014 and 2015. In addition to production growth, other factors contributing to consumption growth also include waste reduction, efficiency enhancement, improvement in sales channels, growth in demand owing to population growth, increased income, and urbanization. International trade also plays an important role (FAO 2016).

Substantial growth in aquatic products consumption provides diverse and rich nutritious food for people around the world, which improves the dietary quality for human beings. In 2013, aquatic products accounted for approximately 17% of the animal protein intake of the global population and 6.7% of the total protein intake. In addition, for more than 3.1 billion people, aquatic products account for nearly 20% of their average daily animal protein intake. In addition to digestible and high-quality protein that includes all of the essential amino acids, aquatic products also provide human beings with essential fats (such as long-chain omega-3 fatty acids) and various vitamins (D, A, and B) and minerals (including calcium, iodine, zinc, iron, and selenium). A small number of aquatic products can significantly improve the nutrition level of people in low-income food-deficit countries and those in the least developed countries, whose dietary pattern is mainly plant based (FAO 2016).

1.3.2.2 Fisheries as Direct Contributors to National Economies

In China, for example, the fisheries gross output value accounted for approximately 1.6% of the overall value from agriculture in 1978, and in 1997, this value reached 10.6%. The per capita income of fishermen increased from RMB 93 in 1978 to RMB 3974 in 1997, 90% higher than the per capita income of farmers; in 2015, tie per capita net income of fishermen even reached RMB 15594.83. Fisheries have become an important industry that promotes the economic prosperity and development of China's rural areas, especially the development of aquaculture, which is one of the effective ways for farmers to live a prosperous and well-off life (FAO 2016).

1.3.2.3 Fisheries as Contributors to Financial Revenue and Foreign Exchange Earnings

International trade plays an important role in capture and aquaculture by providing job opportunities and food, enhancing income and economic growth, and ensuring food and nutrition security. Aquatic products are one of the largest bulk commodities in the global food trade. Approximately 78% of aquatic products participate in international trade competition. For many countries, numerous coastal areas, and regions along rivers, aquatic products export is an economic lifeline. For some island countries, aquatic products export can account for more than 40% of the total value of the merchandise trade, 9% of the total value of global agricultural exports, and 1% of the total value of global merchandise trade. In recent decades, under the drive of aquatic production growth and increased demand, the aquatic products trade volume has significantly increased, while the fisheries sector has also faced a global environment of continuous integration. In addition, service trade related to fisheries is also an important activity (FAO 2016).

Statistically, in 1976, the aquatic products export volume in developing countries accounted for only 37% of the total global trade; however, in 2014, the proportion of the export value increased to 54% and the proportion of the export volume (live weight) increased to 60%. Aquatic products trade has become an important source of foreign exchange earnings for many developing countries. In 2014, the aquatic products export value of developing countries reached US $ 80 billion, and the net value of aquatic products foreign exchange earnings (exports minus imports) reached US $ 42 billion, higher than the total value of other agricultural bulk commodities (such as meat, tobacco, rice, and sugar) (FAO 2016).

1.3.2.4 Fisheries as Contributors to Rural Labor Allocation and Social Employment

In 2014, approximately 56.6 million people were employed in the primary sectors of the fishing industry and aquaculture all over the world, of whom 36% were full-time, 23% part-time, and the rest temporarily employed or unknown. After a long period with an upward trend, employment has remained relatively stable since 2010, while the proportion of people engaged in aquaculture rose from 17% in 1990 to 33% in 2014. In terms of the population employed in fisheries and aquaculture in 2014, 84% were from Asia, 10% from Africa, and 4% from Latin America and the Caribbean. Of the 18 million people engaged in aquaculture, 94% were from Asia. In 2014, women accounted for 19% of the people directly engaged in primary production, but if considering secondary industries (such as processing and trade), women accounted for approximately 50% (FAO 2016).

In addition to the primary production sector, fisheries and aquaculture also provide many people with employment opportunities in subsidiary activities, such as processing, packaging, marketing, equipment manufacturing for processing aquatic products, manufacture of fishing nets and fishing gear, ice making and supply, vessel construction and maintenance, research, and administration. With all of these employment opportunities as well as the large number of family member dependents, an estimated 660–820 million people live on capture and aquaculture, accounting for 10–12% of the global population. In addition, more than 90% of global fishermen are engaged in small-scale fisheries, which play an important role in improving food security and poverty alleviation and prevention (FAO 2016).

1.3.2.5 Fisheries as Contributors to Global Sustainable Development and Aquatic Ecosystems

Fishery resources provide human beings with material functions; more importantly, by fulfilling a role in retaining water and maintaining a balanced ecosystem, they also provide ecological functions. If marine and inland waters (lakes, rivers, and reservoirs) can restore and maintain health and productivity, they could bring huge benefits to human beings. To ensure the sustainability of capture and aquaculture, it is necessary to carry out management of marine, coastal, and inland water ecosystems, including management of habitats and biological resources. The "Blue Growth Initiative" proposed by the FAO not only highlights the ecosystem approach to capture and aquaculture but also proposes promoting the sustainable livelihoods of coastal fishing communities, prioritizing and supporting small-scale capture and aquaculture, and ensuring fair access to trade, markets, social protection, and decent work in all processes of the aquatic products value chain (FAO 2016).

1.3.2.6 Fisheries as Contributors to Other Industries

Aquatic products production belongs to the primary industry but is closely linked to secondary and tertiary industries. Aquatic products production relies on the support of other industries; in return, it provides raw materials for other production sectors, such as food, medicine, feed, light industry, agriculture, and other sectors that use fishery products. Meanwhile, fisheries obtain products provided by sectors involving feed, chemicals, refrigeration, construction, machinery, and shipbuilding (Chen and Zhou 2018).

More importantly, fisheries are first a food supply industry, and a stable food supply is an important foundation for stable national life and stable society; therefore, it is not only important as an economic activity. Considering the widening gap between the world population and increased food production in the future, people should have a better understanding of the importance of fisheries in national economies. Second, most economic activities are currently carried out in large cities, while aquaculture is carried out in small- and medium-sized cities, villages, and islands in coastal areas. Therefore, fisheries play an important role in a particular regional economy. Considering the balanced development of all regions of a national economy, this industry is more important than the significance of its share in a national economy. Meanwhile, fisheries are conducive to rural economic restructuring, rational development and utilization of land resources, and high-tech development, such as new materials, new technologies, new processes, and new equipment.

References

Chen XJ (2014) Fisheries resources economics. China Agriculture Press, Beijing. (in Chinese)
Chen XJ, Zhou YQ (2018) Introduction to fishery. China Science Press, Beijing. (in Chinese)
FAO (2016) The state of world fisheries and aquaculture 2016. Rome

Chapter 2
Review on Global Fisheries

Xinjun Chen, Yinqi Zhou, and Leilei Zou

Abbreviations

APFIC	The Asia-Pacific Fisheries Commission
Cancun Declaration	Cancun International Responsible Fishing Declaration
CCSBT	Commission for the Conservation of Southern Bluefin Tuna
CECAF	Fishery Committee for the Eastern Central Atlantic
CECAF	The Fishery Committee for the Eastern Central Atlantic
COPPESAALC	Commission for Small-Scale and Artisanal Fisheries and Aquaculture of Latin America and the Caribbean
EEZ	Exclusive economic zone
EIFAC	European Inland Fisheries and Aquaculture Advisory Commission
EU	The European Union
FSA	Agreement for the Implementation of the United Nations Convention on the Law of the Sea Relating to the Conservation and Management of Straddling Fish Stocks and Highly Migratory Fish Stocks
GFCM	General Fisheries Commission for the Mediterranean
IATTC	Inter-American Tropical Tuna Commission

X. Chen (✉) · Y. Zhou
College of Marine Sciences, Shanghai Ocean University, Lingang New City, Shanghai, China
e-mail: xjchen@shou.edu.cn; yqzhou@shou.edu.cn

L. Zou
College of Foreign Languages, Shanghai Ocean University, Lingang New City, Shanghai, China
e-mail: llzou@shou.edu.cn

© Science Press & Springer Nature Singapore Pte Ltd. 2020
X. Chen, Y. Zhou (eds.), *Brief Introduction to Fisheries*,
https://doi.org/10.1007/978-981-15-3336-5_2

ICCAT	International Commission for the Conservation of Atlantic Tunas
IFPRI	International Food Policy Research Institute
IOTC	Indian Ocean Tuna Commission
IPHC	International Pacific Halibut Commission
IUU	Fishing illegal, unreported, and unregulated fishing
IWC	International Whaling Commission
NAFO	Northwest Atlantic Fisheries Organization
NEAFC	North-East Atlantic Fisheries Commission
Reykjavik Declaration	Reykjavik Declaration on Responsible Fisheries in the Marine Ecosystem
RFB	Regional Fishery Bodies
RFMO	Regional Fisheries Management Organizations
RFO	Regional fisheries organizations
SEAFO	The South East Atlantic Fisheries Organization
SOFIA	The State of World Fisheries and Aquaculture
SPC	The Pacific Community
SPRFMO	South Pacific Regional Fisheries Management Organization
SWIOFC	The Southwest Indian Ocean Fisheries Commission
The Code of Conduct	The Code of Conduct for Responsible Fishing
UNCLOS	The United Nations Convention on the Law of the Sea
UNFSA	The Agreement for the Implementation of the Provisions of the United Nations Convention on the Law of 10 December 1982 Relating to the Conservation and Management of Straddling Fish Stocks and Highly Migratory Fish Stocks
WCAFC	Western Central Atlantic Fisheries Commission
WCPFC	The Western and Central Pacific Fisheries Commission
WSSDPOI	The World Summit on Sustainable Development Plan of Implementation

2.1 An Overview of the Status of Global Fisheries Development

2.1.1 Overview

With capture fisheries production being relatively static since the late 1980s, aquaculture has been the primary contributor to the impressive growth in the supply of aquatic products for human consumption (Fig. 2.1). Although aquaculture provided only 7% of aquatic products for human consumption in 1974, this share increased to 26% in 1994 and 39% in 2004. China is the primary contributor to this growth and represents more than 60% of world aquaculture production. However, the share of

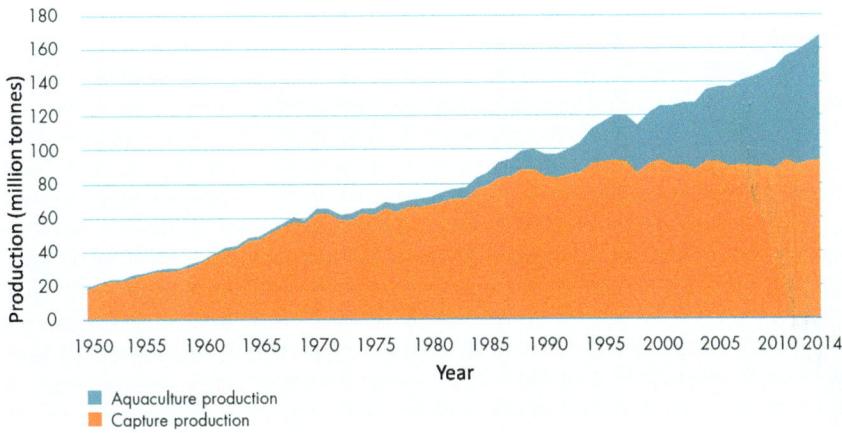

Fig. 2.1 Global capture and aquaculture production (FAO 2016)

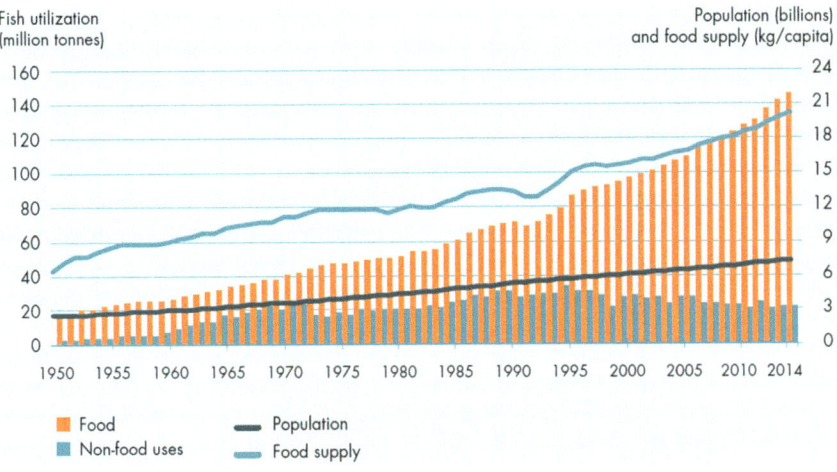

Fig. 2.2 Global fish utilization and supply (FAO 2016)

aquaculture in the overall aquatic products supply in the rest of the world (excluding China) has also more than doubled since 1995.

Growth in the global supply of aquatic products for human consumption has outpaced population growth over the past five decades, increasing at an average annual rate of 3.2% in the period from 1961 to 2013, double that of population growth, resulting in an increasing average per capita availability (Fig. 2.2). Global per capita aquatic products consumption increased from an average of 9.9 kg in the 1960s to 14.4 kg in the 1990s and 19.7 kg in 2013, and this consumption was expected to increase beyond 20 kg (Table 2.1) in 2014 and 2015.

The dramatic increase in consumption of aquatic products has improved the quality of the human diet by providing more diversified and nutritious food for

Table 2.1 Global capture and aquaculture production and utilization (unit: million tons) (FAO 2016)

	2009	2010	2011	2012	2013	2014
Capture						
Inland capture	10.5	11.3	11.1	11.6	11.7	11.9
Marine capture	79.7	77.9	82.6	79.7	81.0	81.5
Subtotal	90.2	89.1	93.7	91.3	92.7	93.4
Aquaculture						
Inland aquaculture	34.3	36.9	38.6	42.0	44.8	47.1
Marine aquaculture	21.4	22.1	23.2	24.4	25.5	26.7
Subtotal	55.7	59.0	61.8	66.5	70.3	73.8
Capture and aquaculture	145.9	148.1	155.5	157.8	163	167.2
Utilization						
Human consumption	123.8	128.1	130.8	136.9	141.5	146.3
Non-food uses	22.0	20.0	24.7	20.9	21.4	20.9
Global population (billions)	6.8	6.9	7.0	7.1	7.2	7.3
Per capita food fish supply (kg)	18.1	18.5	18.6	19.3	19.7	20.1

people around the world. In 2013, aquatic products accounted for approximately 17% of the global population's animal protein intake and 6.7% of all protein consumed. Moreover, aquatic products provide more than 3.1 billion people with almost 20% of their average per capita animal protein intake. Aquatic products are usually high in unsaturated fatty acids and provide health benefits by protecting against cardiovascular disease. Aquatic products also aid fetal and infant development of the brain and nervous system. With their valuable nutritional properties, aquatic products can also play a major role in correcting unbalanced diets and countering obesity when substituted for other food.

2.1.2 Review of the Global Capture Industry

The total global capture production in 2014 was 93.4 million tons, of which 81.5 million tons was from marine waters and 11.9 million tons from inland waters (Table 2.1). For marine fisheries production, China remained the major contributor, followed by Indonesia, the USA and Russia. Anchoveta catches in Peru fell to 2.3 million tons in 2014—half that of the previous year and the lowest since the strong El Niño in 1998—but in 2015, they recovered to more than 3.6 million tons. For the first time since 1998, anchoveta was not the top-ranked catch species as it fell below Alaska pollock. Four highly valuable species (tuna, lobster, shrimp, and cephalopod) registered new record catches in 2014. The total catches of tuna and tuna-like species were almost 7.7 million tons.

The Pacific Northwest remained the most productive area for capture fisheries, followed by the Western Central Pacific, the Northeast Atlantic, and the Eastern Indian Oceans. With the exception of the Northeast Atlantic, these areas have witnessed increased catches compared with the decade average for 2003–2012. The situation in the Mediterranean and Black Sea is alarming, as catches have dropped by one-third since 2007, mainly attributable to reduced landings of small pelagics, such as anchovy and sardine, but with most species also affected.

Global catches in inland waters were approximately 11.9 million tons in 2014, continuing a positive trend that has resulted in a 37% increase over the last decade. In all, 16 countries have annual inland water catches exceeding 200,000 tons, and together, these nations represent 80% of the global total.

2.1.3 Overview of Global Aquaculture

The total global aquaculture production in 2014 amounted to 73.8 million tons, including 49.8 million tons of finfish, 16.1 million tons of mollusks, and 6.9 million tons of crustaceans. China produced 45.5 million tons in 2014, accounting for more than 60% of global aquaculture production. The other major producers were India, Vietnam, Bangladesh, and Egypt. In addition, aquaculture included 27.3 million tons of aquatic plant production, with seaweed as the dominant product, the production of which underwent rapid growth, and approximately 50 countries were engaged in seaweed aquaculture. More importantly, in terms of food security and environment, approximately half of global aquaculture production, including aquatic animals and plants, comes from non-fed species, which include silver and bighead carps, filter-feeding animals (e.g., bivalve mollusks) and seaweed. However, production growth has been faster for fed species than for non-fed species.

2.1.4 Review of the Global Fisheries Fleet and Employment

An estimated 56.6 million people were engaged in the primary sector of capture fisheries and aquaculture in 2014, of whom 36% were engaged full time, 23% part time, and the remainder either occasional fishermen or of unspecified status. After a constant upward trend, the fishing population has remained relatively stable since 2010, while the proportion of fishermen engaged in aquaculture increased from 17% in 1990 to 33% in 2014. In 2014, 84% of the global population engaged in the capture and aquaculture sector was in Asia, followed by Africa (10%), and Latin America and the Caribbean (4%). Of the 18 million people engaged in aquaculture, 94% were in Asia. Women accounted for 19% of fishers who were directly engaged in this primary sector in 2014, but when the secondary sector (e.g., processing and trading) is included, women make up approximately half of the workforce.

The total number of fishing vessels in the world was estimated at approximately 4.6 million in 2014, and the most important contributor was Asia, where there were 3.5 million vessels, accounting for 75% of the global fleet, followed by Africa (15%), Latin America and the Caribbean (6%), North America (2%), and Europe

(2%). Globally, 64% of reported fishing vessels were engine-powered in 2014, of which 80% were in Asia, with the remaining regions all under 10%. In 2014, approximately 85% of the world's engine-powered fishing vessels were less than 12 m in length, and approximately 64 thousand fishing vessels longer than 24 m in length were operating in marine waters.

2.1.5 Status of Global Marine Fishery Resources

The status of global marine fishery resources has not improved, despite notable progress in some areas. Based on an FAO analysis of assessed commercial fish species, the share of fish species within biologically sustainable levels decreased from 90% in 1974 to 68.6% in 2013. Thus, 31.4% of fish species were estimated to be at a biologically unsustainable level and, therefore, overfished. Of the total number of stocks assessed in 2013, fully fished species accounted for 58.1% and underfished stocks 10.5%. Since 1990, the number of species fished at unsustainable levels has continued to increase, albeit at a slower speed. The ten most-productive species accounted for approximately 27% of the global marine capture production in 2013. However, most of these species are fully fished with no potential for increased production; the remainder are overfished with increased production only possible after successful stock restoration.

2.1.6 Review of the Processing and Utilization of Global Aquatic Products

The share of global aquatic products utilized for direct human consumption has significantly increased in recent decades, up from 67% in the 1960s to 87% in 2014, that is, more than 146 million tons in 2014. The remaining 21 million tons was destined for non-food products, of which 76% was for fishmeal and fish oil in 2014 and the rest largely utilized for a variety of purposes including as raw materials for direct feeding in aquaculture. Impressively, the utilization of by-products is becoming an important industry, with a growing focus on processing in a controlled, safe, and hygienic way, thereby reducing waste.

In 2014, 46% (67 million tons) of aquatic products were consumed directly by human beings as live, fresh, or chilled, the most preferred and highly priced forms at markets. The rest of the food products were in different processed forms, with approximately 12% (17 million tons) in dried, salted, smoked, or other cured forms, 13% (19 million tons) in prepared and preserved forms, and 30% (approximately 44 million tons) in frozen form. Freezing is the main method of processing aquatic products for human consumption, and it accounted for 55% of the total processed aquatic products for human consumption and 26% of total aquatic production in 2014.

Fishmeal and fish oil are still considered the most nutritious and digestible ingredients for farmed fish feed. To offset their high prices, as feed demand increases, the amount of fishmeal and fish oil used in compound feeds for aquaculture has shown a clear downward trend, as they are more selectively used as strategic ingredients at lower concentrations and for specific stages of production, particularly for hatchery, broodstock, and finishing diets.

2.1.7 Review of Global Fisheries Trade

International trade plays a major role in the fisheries and aquaculture sector as a job creator, food supplier, income generator, and contributor to economic growth and development as well as food and nutrition security. Aquatic products represent one of the most traded segments of the global food sector, with approximately 78% of seafood products estimated to be exposed to international trade competition. For many countries and numerous coastal and riverine regions, aquatic products exports are essential to their economies, accounting for more than 40% of the total value of traded commodities in some island countries and globally representing more than 9% of total agricultural exports and 1% of global merchandise trade in value terms. Aquatic products trade has expanded considerably in recent decades, fueled by growing fishery production and high demand, with the fisheries sector operating in an increasingly globalized environment. In addition, fisheries-related trade in services is developing as well.

China is the main producer and largest exporter of aquatic products. China is also a major importer due to aquatic product outsourcing for processing from other countries as well as growing domestic consumption of products produced abroad. However, in 2015, after years of constant increases, China's aquatic products trade experienced a slowdown with a reduction in its processing sector. Norway, the second largest exporter, posted record export values in 2015. In 2014, Vietnam became the third largest exporter, overtaking Thailand, which has experienced a substantial decline in exports since 2013 mainly for reduced shrimp production due to disease problems. In 2014 and 2015, the European Union (Member Organization; referred to as the EU) was by far the largest single market for imported fish, followed by the USA and Japan.

Developing economies, whose exports represented only 37% of world trade in 1976, saw their share rise to 54% of the total aquatic products export value and 60% of quantity (live weight) in 2014. Aquatic products trade represents a significant source of foreign currency earnings for many developing countries, in addition to its important role in income generation, employment, food security, and nutrition. In 2014, aquatic products exports from developing countries were valued at USD 80 billion, and their aquatic products net export revenues (exports minus imports) reached USD 42 billion, higher than other major agricultural commodities (such as meat, tobacco, rice, and sugar) combined.

2.2   Division of Main Fishery Resources and Fishing Areas in the World

At the global level, marine organisms are rich both in species and quantity. There are approximately 200,000 species of marine organisms, including 180,000 marine animals, more than 6000 marine plants, more than 500 marine fungi, and more than 12,000 marine protozoa. Marine animals can be divided into 20,000 fish species, 30,000 crustacean species, and 100,000 mollusk species. The total biomass is approximately 32.5 billion tons, including 21.5 billion tons of zooplankton, 1 billion tons of swimming animals, 10 billion tons of benthic animals, and 1.7 billion tons of marine plants. Living marine resources weigh more than twice that of land-based animals (10 billion tons).

Global fisheries are an important source of human consumption. This term refers to animals and plants that inhabit and reproduce in water and that have economic, exploitation, and utilization value. In the *Yearbook of Fishery Statistics* published by the FAO in 2003, 1223 marine and freshwater aquatic animals species are listed, and they fall under 7 categories, which include 900 fish species, 122 crustacean species, 97 mollusk species, 67 mammal species, 19 amphibian and reptile species, 18 aquatic invertebrate species, and 21 aquatic plant species. The aquatic plants can be further divided into *Cyanophyceae*, *Chlorophyceae*, *Phaeophyceae*, *Rhodophyceae*, and *Angiospermae*.

2.2.1   Classification by Habitat Layer

Aquatic economic animals such as fish have different habits and physiological needs, which are reflected by different habitat preferences in water. However, there are also seasonal changes in habitat layers due to the need for reproduction, feeding, and wintering. There is a significant change in habitat layer between day and night due to the influence of light. According to habitat layer, fishery resources can be classified as follows.

2.2.1.1   Demersal Species

Demersal species refer to those species living at or near the bottom of the water body, which include finfish, crab, and mollusk. For example, the life cycle of demersal fish is generally longer than other species, with some living for more than 30 years. Once overfishing causes resource decline, it is difficult to recover these resources. Major demersal fish are as follows.

Fig. 2.3 Schematic diagram of external morphology of Pacific pollock

Fig. 2.4 Schematic diagram of external morphology of Atlantic cod

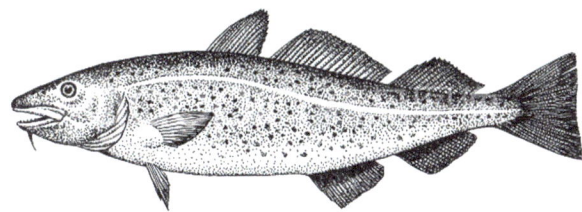

Pacific Pollack (*Theragra chalcogramma*)

This species is slightly slender and has a large head with three dorsal fins, two anal fins, large eyes, a prominent chin, and short whiskers on the jaw. Its body color is olive, with lighter coloring on the belly and many small spots on the body. Pacific pollock meat is white (Fig. 2.3).

Pacific pollock is distributed in the North Pacific, especially in the Bering Sea, the Okhotsk Sea, and the Sea of Japan, as well as the open seas of California and northern Alaska in the USA. Pacific pollock has a powerful air bladder. This fish usually inhabits the deep-sea floor hundreds of meters underwater, but it can rise or sink quickly. This species moves in large, size-structured aggregations. The countries with the largest catches are Russian, Japan, South Korea, China, and Poland, among others. Pacific pollock is captured by large trawls, which can catch several tons per time. Pacific pollock can be made into fish fillets or surimi after removal of the head, organs, bones, and skin and can be frozen for storage. Additionally, the waste can be made into fishmeal and fish oil.

Atlantic Cod (*Gadus morhua*)

The Atlantic cod body has an elliptical cross section with three distinct dorsal fins, two anal fins, and an almost square caudal fin. Cod coloring varies from brownish to greenish with the surrounding environment. This species has dots on its back. Usually, cod is 0.9 m in length and weighs between 4.5 and 11.3 kg (Fig. 2.4).

The Atlantic cod is distributed on both sides of the Atlantic Ocean, along the North American coast, further northward to Greenland, Davis Strait, Hudson Strait,

Fig. 2.5 Schematic
diagram of external
morphology of
Argentine hake

Fig. 2.6 Schematic
diagram of external
morphology of Peruvian
anchovy

and southward to Cape Hatteras. However, in Europe, this species is found on the islands of Isla Mujeres and Spitsbergen and from Jan Mayen to Vizcaya in Norway, and it is also distributed near Iceland and the Faroe Islands. This fish usually inhabits areas ranging from coastal shallow water to depths up to 450 m. The countries with the largest catches are Russia, Norway, Denmark, and Iceland, among others, and this species is widely captured by trawling, drifting traps, and angling. Most catches are sold fresh or frozen, while some are sold marinated or dried. The waste is made into fishmeal and fish oil.

Argentine Hake (*Merluccius hubbsi*)

Argentine hake has no whiskers on its jaw (Fig. 2.5) and is an important economic fish species in the southwestern Atlantic Ocean.

Argentine hake inhabits mainly continental shelf waters along the eastern coast from approximately 28°S to 54°S off southern South America, ranging from 50 to 500 m in depth. Argentina is one of the countries with the largest catches. This fish is widely captured by trawling.

2.2.1.2 Pelagic Species

Pelagic species refer to those species inhabiting in the pelagic zone of water, neither close to the bottom nor near the shore, most of which are fish. Generally, these species grow rapidly and have high productivity and relatively short life cycles. Pelagic fish, such as the Peruvian anchovy (*Engraulis ringens*), are long and narrow in shape and shiny blue or green in color and have a round body cross section (Fig. 2.6).

The Peruvian anchovy is distributed in the outer waters of Peru and Chile in the Southeast Pacific Ocean, between 5°S–43°S and 82°W–69°W. The strength and distribution of the Peruvian current has an important impact on its distribution. This fish is generally distributed in an area approximately 80 km from the coast, in water temperatures ranging from 13 °C to 23 °C. Its size is typically approximately 20 cm long. This species reportedly has a longevity of approximately 3 years. The Peruvian anchovy is widely captured by trawling.

2.2.1.3 Highly Migratory Fish Species

Highly migratory fish are the main targets of oceanic high seas fisheries, and they migrate in the Pacific Ocean, Atlantic Ocean, and Indian Ocean. According to the *United Nations Convention on the Law of the Sea* (referred to as UNCLOS), these species include tuna, skipjack tuna, freshwater bream, spearfish, billfish, sailfish, Pacific saury, oceanic shark, and whale.

Tuna is a sleek and streamlined fish, adapted for high-speed movement. This species is plump and has a circular cross section. The caudal peduncle is quite thin, with a pair of bulges attached on both sides. This sleek and streamlined fish has thick skin, adapted for intense disturbance. Tuna is covered with scales, except for the head. The central part of the back of the body is generally dark blue, while some fish are slightly lighter. The ventral part is generally white.

The main tuna species are yellowfin tuna, bigeye tuna, bluefin tuna, albacore tuna, and skipjack tuna (Figs. 2.7, 2.8, 2.9, 2.10, 2.11, and 2.12). They are widely but sparsely distributed in the area of 45°S–45°N.

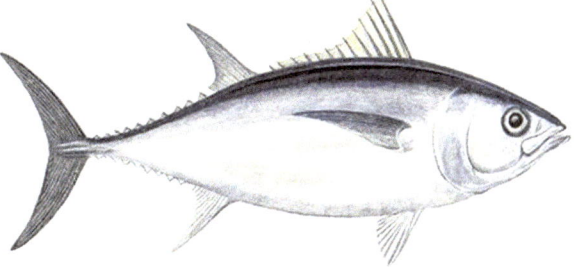

Fig. 2.7 Schematic diagram of external morphology of southern bluefin tuna

Fig. 2.8 Schematic diagram of external morphology of longfin tuna

Fig. 2.9 Schematic diagram of external morphology of bigeye tuna

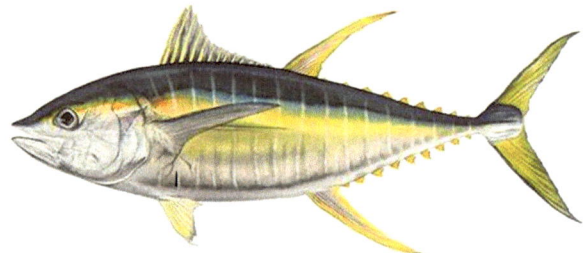

Fig. 2.10 Schematic diagram of external morphology of yellowfin tuna

Fig. 2.11 Schematic diagram of external morphology of skipjack tuna

2.2.1.4 Anadromous Fish Species

Anadromous fish migrate from the ocean to inland rivers, where they lay eggs and die after spawning, while juveniles migrate to the ocean, where they grow up and, 4 years later, migrate back to the original river for spawning. Here are some examples:

Atlantic salmon (*Salmo salar*) is slightly different from the Pacific salmon. Not all of them die after spawning, and some can lay eggs up to three times in their life. The body length and weight are reported to be 150 cm (male), 120 cm (female), and 46.8 kg, respectively. Its maximum longevity is 13 years. Its favorite habitat is cold water ranging from 2 °C to 9 °C, mainly in the area of 72°N–37°N and 77°W–61°E. Its habitat depth is 0–210 m. Generally, Atlantic salmon fry live in freshwater for 1–6 years and then migrate to the ocean, where they live for 1–4 years and finally migrate back to the river again for spawning. This fish grows faster in the ocean than in the river (Fig. 2.13).

Catadromous fish spend most of their lives in freshwater and then migrate to the ocean for spawning. The river eel is a type of catadromous fish (Fig. 2.14).

Fig. 2.12 Schematic diagram of external morphology of swordfish

Fig. 2.13 Schematic diagram of external morphology of Atlantic salmon

Fig. 2.14 Schematic diagram of external morphology of river eel

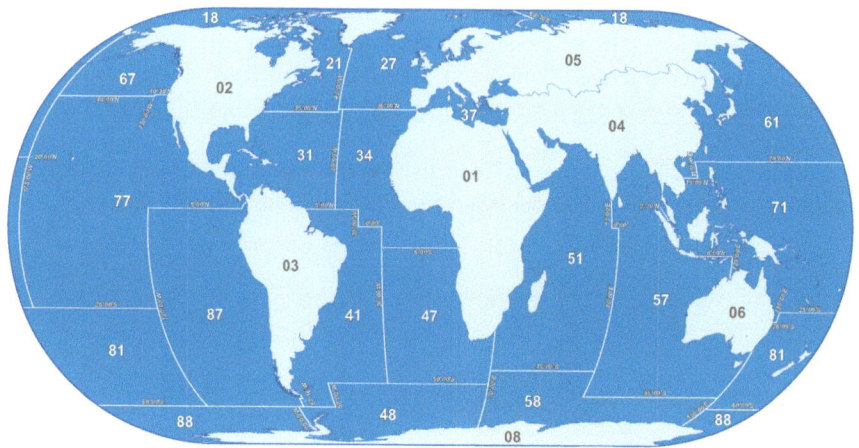

Fig. 2.15 Schematic diagram of continental and oceanic fishing areas (FAO 2016)

2.2.2 Classification of Global Fishing Areas

To facilitate fishery statistics, the FAO divides the continents and oceans into different fishing areas, and the fishing area is numbered in turn (Fig. 2.15).

2.2.2.1 Continental Fishing Areas

Continental fishing areas are numbered according to the continent where the fishing areas are located, and they are listed as follows: Africa–"01"; North America– "02"; South America– "03"; Asia– "04"; Europe– "05"; Oceania– "06"; and Antarctica– "08".

2.2.2.2 Oceanic Fishing Areas

Oceanic fishing areas are numbered according to the ocean where fishing areas are located, and they are listed as follows:

The Arctic Ocean– "18"

The Atlantic Ocean:

The Northwest Atlantic Ocean– "21"; the Northeast Atlantic Ocean– "27"; the Western Central Atlantic Ocean– "31"; the Eastern Central Atlantic Ocean– "34"; the Southwest Atlantic Ocean– "41"; and the Southeast Atlantic Ocean– "47"

The Indian Ocean:

The Western Indian Ocean– "51"; the Eastern Indian Ocean– "57"

The Pacific Ocean:

The Northwest Pacific Ocean– "61"; the Northeast Pacific Ocean– "67"; the Western Central Pacific Ocean– "71"; the Eastern Central Pacific Ocean– "77"; the Southwest Pacific Ocean– "81"; and the Southeast Pacific Ocean– "47"

Table 2.2 Surface freshwater distribution in continents and countries

	Surface area (km^2)							Total	Percentage (%)
	Lake	Reservoir	River	Floodplains	Flooded forest	Peatland	Wetland		
Asia	898,000	80,000	141,000	1,292,000	57,000	491,000	357,000	3,316,000	42
South America	90,000	47,000	108,000	422,000	860,000	–	2800	1,529,800	20
North America	861,000	69,000	58,000	18,000	57,000	205,000	26,000	1,294,000	17
Africa	223,000	34,000	45,000	694,000	179,000	–	187,000	1,362,000	17
Europe	101,000	14,000	5000	53,000	–	13,000	500	186,500	2
Australia	8000	4000	500	–	–	–	112,000	124,500	2
Oceania	5000	1000	1000	6000	–	–	100	13,100	0
Total	2,186,000	249,000	358,500	2,485,000	1,153,000	709,000	685,400	7,825,900	100

The Southern Ocean:

The South Atlantic Ocean– "48"; the Southern Indian Ocean– "58"; and the South Pacific Ocean– "88"

2.2.3 Living Resources and Fishery Types in Global Inland Water

2.2.3.1 Areas of Inland Water

The total size of lakes, reservoirs, and wetlands in the world is approximately 7.8 million km^2 (Table 2.2), and these water bodies are important for inland fisheries. A relatively high proportion of land in Southeast Asia, North America, East Africa, central and West Africa, northern Asia, Europe, and South America is covered by surface water.

Sourced from: Lehner B, Doell P (2004)

2.2.3.2 Types of Inland Water Fisheries

Inland water fisheries are diverse in type, and they involve a wide range of fishing techniques, ranging from simple hand-held nets to small trawls or purse seines operated by commercial fishing vessels. Inland water fisheries involve commercial fisheries, industrialized fisheries, small-scale fisheries, and recreational fisheries, each with different economic and social structures.

Commercial and Industrialized Inland Fisheries

The main motivation for many fishermen, including fishermen engaged in small-scale fisheries, is income. Since modern small-scale fisheries can also be economically efficient and provide the international market with high value-added products, they are not limited to commercial and industrialized fisheries.

Commercial inland fisheries produce large quantities of aquatic products in certain selected areas. These products often require specialized preservation and sale, which usually involves high-capital investment in fishing gear as well as abundant professional labor. Commercial fisheries often emerge where fishing is controlled, and significant investments (financial, human resources, and/or net construction) are justified by such favorable conditions as resource availability and market access. Key fishing sites or opportunities are usually allocated by auctions through a licensing system. Commercial and industrialized inland fisheries include lake fisheries in developed countries, the Great Lakes fisheries in Africa, and sturgeon fisheries in the Caspian Sea. However, some commercial and industrial river fisheries in Southeast Asia are also very impressive, such as fisheries in the "fishing area" and net fisheries in Cambodia, as well as the fishing hotels and reservoir sales concessions in Myanmar. In Latin America, equally impressive are the migratory catfish fisheries in the Amazon and the industrialized Brazilian bream fisheries in the Platte River.

Small-Scale Inland Fisheries

Small-scale inland fisheries are active and evolving, developing fishery resources by taking advantage of their intensive labor, processing, and marketing techniques. Small-scale inland fisheries are carried out by both full-time and part-time fishermen, supplying fish and fishery products to the local market, or occasionally to the domestic market as well. The composition of fishermen occasionally engaged in fishery operation is complicated. Given the opportunity, they are engaged in fishery operation to earn money to support their families. The number of occasionally engaged fishermen is, more often than not, more than that of full-time and part-time fishermen. Since surplus products in the smallest-scale fisheries are sold or exchanged for other products or services, pure subsistence fisheries are rare. Subsistence fishing means more family-centered activities than commercial activities. Even if the captured fish is not for sale, it is still valuable for local consumption because such products contribute to the welfare and food security of a household, the local area, or the regional area.

Recreational Fisheries

Recreational fisheries involve fishing for pleasure, for competition, or for self-consumption. Recreational fishing is a popular activity and form of entertainment in many developed (e.g., in some western European countries or areas, Australia, Canada, New Zealand, and the USA) as well as in some developing countries (e.g., Argentina, Botswana, Brazil, Chile, Mexico, South Africa, and Thailand; among these countries, some are newly emerging developing countries). As the definition indicates, recreational fisheries do not fall under the category of commercial activity, and catches are not usually for sale. Catches may be released back into the water, kept as souvenirs, eaten, or sold, but eating and selling are not the main purpose of recreational fishing. However, recreational fisheries can greatly promote local and national economies through creating employment opportunities in supporting industries.

2.3 Evolution and Structural Change in Global Fisheries

2.3.1 Evolution of Global Fisheries

The marine capture industry used to be the main contributor to global fisheries. Since the 1990s, however, much priority has been given to aquaculture development. As for the total volume of global fisheries, the proportion of annual marine capture production has been gradually declining. The international community has also recognized that it is not marine capture but aquaculture that has a promising future and development potential.

The capture industry has developed alongside socioeconomic development and the advancement of science and technology. For the convenience of better analysis and research, the evolution of global fisheries production after the end of the Second World War can be divided into the following stages: the recovery and development stage in the 1950s, the rapid development stage in the 1960s, the stagnation stage in the 1970s, the development of high seas fisheries and aquaculture in the 1980s, and the management and industry structure adjustment stage since the 1990s.

2.3.1.1 Recovery and Development Stage in the 1950s

During the Second World War, which ended in 1945, a large number of fishing vessels in coastal states were damaged and unable to conduct marine fishing. As a result, the marine fishery resources were relatively rich at that time. After the Second World War, whether it was a victorious country or a defeated country, most countries were eager to solve the problem of grain and food shortages. Coastal states actively restored and developed coastal and inshore fishing as long as conditions permitted. Investment in the capture industry was relatively less than that in agriculture and animal husbandry, while it was paid back very efficiently. The total global fisheries production reached 21.1 million tons in 1950, exceeding the record of 18 million tons in 1938 before the Second World War, and increased to 36.9 million tons in 1959. The average annual growth in global fisheries production during the decade of the 1950s was 1.58 million tons, with an annual growth rate of 7.48%.

2.3.1.2 Rapid Development Stage in the 1960s

In the 1960s, science and technology development greatly promoted fisheries productivity. First of all, the development of the shipbuilding industry improved the seaworthiness and fishing fitness of fishing vessels and provided technical support for the construction of large fishing vessels. The vessels were also equipped with processing equipment, which greatly expanded the fishing area by allowing catches to be directly processed on board the vessel. Second, the horizontal and vertical fish detection instruments invented based on the sonar technology during the Second World War could directly detect the water layer inhabited by fish in the sea and estimate the number of fish. Finally, synthetic fiber materials were commonly used instead of natural fiber materials such as cotton and linen, which greatly improved the speed of fishing nets, hooks, and ropes and prolonged their life span as well. In this way, the countries with rapid fisheries development were actively engaged in extending their fishing fleets across the ocean, exploring new fishing areas and resources. By 1999, the total global fisheries production reached 7.09 million tons. The average annual growth in global fisheries production during the decade of the 1960s was 2.25 million tons, with an annual growth rate of 4.24%.

2.3.1.3 Stagnation Stage in the 1970s

The deep-sea fisheries conducted by countries with rapid fisheries development were actually carried out in the inshore coastal waters of other countries in the 1960s. At that time, the territorial waters of coastal states were only 3 nautical mile in general, and the area beyond 3 nautical mile was the high seas. According to the traditional Law of the Sea, each country was entitled to fishing freedom in the high seas, which were free of coastal state jurisdiction. Since the late 1960s, to prevent developed countries with advanced scientific and technological capabilities from exploiting the mineral resources of the high seas and the inshore fishery resources of coastal states, a vast number of developing countries requested that the United Nations convene a third United Nations Conference on the Law of the Sea to develop a new convention. The United Nations General decided in 1971 to convene the Third United Nations Conference on the Law of the Sea in 1973. After rounds of negotiations, the UNCLOS was finally signed in 1982. During the negotiating period, many coastal states from Asia, Africa, and Latin America released independent declarations of jurisdiction over the waters within 30, 50, 70, and 110 nautical mile and the farthest at 200 nautical mile beyond the coastline. A coastal state could impose a penalty of seizure, fine, or sentence on a foreign fishing vessel that was not allowed to fish in the waters under its jurisdiction, which imposed some restrictions on deep-sea fishing states. At the same time, Peruvian anchovy resources in Peru in the early 1970s were greatly affected by the Peruvian current (the temperate cold current) moving from south to north and El Niño in the Pacific Ocean, with production falling from 12 million tons in 1970 to 2 million tons in 1971, which directly caused the fluctuation in the total global fisheries production. The total global fisheries production in the 1970s stagnated at approximately 70 million tons.

2.3.1.4 Development Stage of High Seas Fisheries and Aquaculture in the 1980s

The Third United Nations Conference on the Law of the Sea adopted the UNCLOS in 1982, providing coastal states with the right to establish an exclusive economic zone (referred to as EEZ) that does not exceed 200 nautical mile from the territorial sea baseline, while providing other countries with the right to engage in fishing activities in an EEZ by complying with relevant regulations, paying the fishing access fee, and so on. As a result, the conflict between coastal states and deep-sea fishing states was alleviated.

Considering the restrictions on fishing in the EEZ of a coastal state, the deep-sea fishing states thus turned to the high seas for fishing, which promoted the development of fisheries production. At the same time, since the mid-to-late 1980s, China has implemented a fisheries production policy of "aquaculture-oriented fisheries policy with coordinated priorities of aquaculture, capture and processing; adapting to local conditions where different priorities are highlighted." With this policy, high priority was given to inland aquaculture, contributing to the unprecedented growth in China's fisheries production, promoting adjustment of the global fisheries industry structure. Aquaculture production started receiving global attention.

Accordingly, in the 1980s, the total global fisheries production exited the stagnation stage of the 1970s and significantly increased. Production exceeded 80 million tons in 1984, 90 million tons in 1986, and 100 million tons in 1989.

2.3.1.5 Fisheries Management and Industry Structure Adjustment Stage Since the 1990s

During the 3 years from 1990 to 1992, the annual global marine capture productions were lower than that in 1989. With excessive development of traditional inshore economic fish and high seas fishery resources, there were indications of the exhaustion of some demersal fish resources, which aroused the attention of the international community. The *Agenda 21* adopted at the United Nations Conference on Environment and Development, held in Rio de Janeiro, Brazil, in 1992, put forward a new concept of sustainable development and some suggestions on the conservation, utilization, and exploitation of living marine resources. Accordingly, the FAO convened a meeting of ministers in Cancun, Mexico, to discuss the responsible fisheries. The United Nations General Assembly decided to ban the use of large-scale spur nets in the ocean from January 1st, 1993. In August 1995, the United Nations Fisheries Conference adopted the *Agreement for the Implementation of the United Nations Convention on the Law of the Sea Relating to the Conservation and Management of Straddling Fish Stocks and Highly Migratory Fish Stocks* (referred to as FSA), which specified how to implement and improve the conservation and management of straddling fish stocks and highly migratory fish stocks in accordance with the UNCLOS. Moreover, the FAO adopted the *Code of Conduct for Responsible Fisheries* in 1995. In general, marine fisheries shifted the focus from fishery resources development to fisheries resource management. In addition, the global fisheries industry structure was adjusted, and the development of aquaculture was becoming increasingly important.

Global fisheries production (including aquatic plant production) has been on the rise since 1990, increasing from 122.7 billion tons in 1990 to 199.7 million tons in 2015. The main source of the growth has been aquaculture. Capture production has been stable at 84–95 million tons, while aquaculture production (including aquatic plant production) continued to increase from 16.85 million tons in 1990 to 106 million tons in 2015. In 2013, aquaculture production (including aquatic plant production), for the first time, exceeded capture production.

2.3.2 Structural Changes in Global Fisheries Production

The structure of global fisheries production mainly illustrates the structure and trends of inland and marine fisheries, aquaculture and capture, the main fishing targets, and the structure and trends of fisheries production in important fishing states.

2.3.2.1 Inland Fisheries and Marine Fisheries

Both inland fisheries and marine fisheries can be divided into capture and aquaculture. Although inland fisheries are lower than marine fisheries in terms of production, inland fisheries are growing faster than marine fisheries. According to statistics on global fisheries and on inland fisheries and marine fisheries (including aquatic plant production) from 1990 to 2015 in Table 2.3, for inland fisheries, the productions in 1995 and 2000 were 20.8859 million tons and 27.3685 million tons, which increased by 48.04% and 93.98%, respectively, compared with 1990. Accordingly, the productions in 2005, 2010, and 2015 increased to 35.6081 million tons, 48.0226 million tons, and 60.3204 million tons, which increased by 30.11%, 75.47%, and 120.40%, respectively, compared with 2000. The productions of marine fisheries in 1995 and 2000 were 104.0429 million tons and 109.1199 million tons, which were 17.21% and 22.93% higher than that in 1990, respectively. Additionally, the productions in 2005, 2010, and 2015 reached 115.91 million tons, 118.8535 million tons, and 139.4208 million tons, which were 6.22%, 8.92%, and 27.77% higher than that in 2000, respectively.

In light of the proportion of the production of inland fisheries and marine fisheries from 1990 to 2015, the proportion of inland fisheries to global fisheries production increased from 13.71% in 1990 to 20.05% in 2000 and further increased to 28.78% in 2010 and 30.20% in 2015. However, the proportion of marine fisheries production to global fisheries production decreased year by year, from 86.29% in 1990 to 79.95% in 2000 and further to 71.22% in 2010 and 69.80% in 2015 (Table 2.4).

Table 2.3 Global inland fisheries and marine fisheries productions from 1990 to 2015 (unit: 10,000 tons)

	1990	1995	2000	2005	2010	2015
Inland capture	644.35	728.66	858.66	943.39	1103.61	1146.95
Inland aquaculture	766.51	1359.93	1878.19	2617.42	3698.65	4885.09
Subtotal	1410.86	2088.59	2736.85	3560.81	4802.26	6032.04
Marine capture	7958.08	8640.99	8617.72	8426.40	7781.99	8226.75
Marine aquaculture	918.53	1763.30	2294.27	3164.60	4103.36	5715.33
Subtotal	8876.61	10404.29	10911.99	11591.00	11885.35	13942.08
Total	10287.47	12492.88	13648.84	15151.81	16687.61	19974.12

Table 2.4 Proportion of global inland fisheries and marine fisheries production to global fisheries production from 1990 to 2015 (%)

	1990	1995	2000	2005	2010	2015
Inland fisheries	13.71	16.72	20.05	23.50	28.78	30.20
Marine fisheries	86.29	83.28	79.95	76.50	71.22	69.80

Table 2.5 Global aquaculture and capture productions from 1990 to 2015 (unit: 10,000 tons)

	1990	1995	2000	2005	2010	2015
Inland capture	644.35	728.66	858.66	943.39	1103.61	1146.95
Marine capture	7958.08	8640.99	8617.72	8426.40	7781.99	8226.75
Subtotal	8602.43	9369.65	9476.38	9369.79	8885.60	9373.70
Inland aquaculture	766.51	1359.93	1878.19	2617.42	3698.65	4885.09
Marine aquaculture	918.53	1763.30	2294.27	3164.60	4103.36	5715.33
Subtotal	1685.04	3123.23	4172.46	5782.02	7802.01	10600.42
Total	10287.47	12492.88	13648.84	15151.81	16687.61	19974.12

2.3.2.2 Aquaculture and Capture

Both aquaculture and capture can be divided into inland fisheries and marine fisheries. Over time, the production of aquaculture (including aquatic plants) has been higher than that of capture, but the growth rate of aquaculture was much higher than that of capture from 1990 to 2015 (Table 2.5). Before 2013, capture production was higher than aquaculture but less stable. Inland capture production has constantly increased year after year, except for a minor drop in 2012. However, marine fisheries production has fluctuated, which, in light of FAO reports, is related to fluctuations in Peruvian anchovy production due to El Niño. The anchovy production in Peru was only 1.7 million tons in 1998 and reached 11.3 million tons in 2000. In the following years, production fluctuated and reached 10.7 million tons in 2004. The total global fisheries production in 2015 was lower than the historical production record (93.7369 million tons), but it was approximately 5 million tons higher than that in 2010, and this increase mainly originated from the marine capture industry. However, inland fisheries production continued to increase slightly, reaching 11.4695 million tons in 2015, the historical record to date.

In an analysis of the growth rate in aquaculture production (Table 2.5), aquaculture production was found to increase by 85.35% in 1995 and 147.62% in 2000 compared to 1990. Accordingly, compared with 2000, this rate increased by 38.58%, 86.99%, and 154.06% in 2005, 2010, and 2015, respectively. Inland fisheries production in 1995 and 2000 increased by 77.42% and 145.03%, respectively, compared with 1990, while increasing in 2005, 2010, and 2015 by 39.36%, 96.93%, and 160.01% compared with 2000, respectively. Marine aquaculture production in 1995 and 2000 increased by 91.97% and 149.78%, respectively, compared with 1990, while in 2005, 2010, and 2015 increased by 37.93%, 78.95%, and 149.11% compared with 2000, respectively. The growth rate of capture production was much lower than that of aquaculture production. The capture productions in

Table 2.6 Proportion of aquaculture and capture production to global fisheries production from 1990 to 2015 (%)

	1990	1995	2000	2005	2010	2015
Capture	83.62	75.00	69.43	61.84	53.25	46.93
Aquaculture	16.38	25.00	30.57	38.16	46.75	53.07

1995 and 2000 increased by 8.92% and 10.16%, respectively, compared with 1990. Compared with 2000, productions in 2005, 2010, and 2015 decreased by 1.13%, 6.24%, and 1.08%, respectively.

From 1990 to 2015, aquaculture production increased from 16.38% in 1990 to 30.57% in 2000 and increased to 46.75% in 2010 and 53.07% in 2015. Meanwhile, capture production decreased from 83.62% in 1990 to 69.43% in 2000 and further decreased to 53.25% in 2010 and 46.93% in 2015 (Table 2.6). As shown in Table 2.6, in global fisheries production (including aquatic plant production), aquaculture production has exceeded capture production, which highlights the significant role of aquaculture in global fisheries.

2.4   Global Marine Capture Industry

2.4.1   Major Marine Fishing States

For a long period prior to the mid-1980s, the top three marine fishing states in the world were Japan, USSR (Russia since 1991), and China. Due to the high fluctuations in anchovy production, Peru ranked highly in some years. However, Japan and Russia have markedly declined in their fisheries production since the 1990s.

Japan: Due to the lack of domestic labor and rising oil prices, Japan's fisheries production cost has increased. Its fisheries development is also restricted by the establishment of EEZs of coastal states and the increasingly strict management of high seas fisheries. Based on the above situation, Japan's marine capture industry is tending to shrink, and its production continues to decline. However, Japan has a traditional diet with fish as an important food. Therefore, Japan has changed from a fish producing state to a fish trading state, importing fish to meet its domestic needs. The highest Japanese production was 12 million tons in 1988, followed by 4.4 million tons in 2002, 4.8 million tons in 2004, 3.61 million tons in 2014, and only 3.43 million tons in 2015.

Russia: Since the 1950s, the USSR has vigorously developed deep-sea marine fisheries. Russian deep-sea fishing fleets were large scale, with large trawl processing vessels as the dominant type, and Russia fished in the Northeast Atlantic, the Eastern Central Atlantic, and the Southeast Atlantic, the Western Indian Ocean, and the Pacific Northwest. Russian annual production reached 7–8 million tons. After the disintegration of the USSR, the state-owned enterprises collapsed, and the annual production continued to decline, at only 3.7 million tons in 1994,

Table 2.7 Top 10 marine fishing states in the World in 2002, 2004, and 2015 (unit: million tons)

2002		2004		2015	
Country	Production	Country	Production	Country	Production
1. China	16.6	1. China	16.9	1. China	15.31
2. Peru	8.8	2. Peru	9.6	2. Indonesia	6.03
3. USA	4.9	3. USA	5.0	3. USA	5.02
4. Indonesia	4.5	4. Chile	4.9	4. Peru	4.79
5. Japan	4.4	5. Indonesia	4.8	5. Russia	4.17
6. Chile	4.3	6. Japan	4.4	6. India	3.50
7. India	3.8	7. India	3.6	7. Japan	3.43
8. Russia	3.2	8. Russia	2.9	8. Vietnam	2.61
9. Thailand	2.9	9. Thailand	2.8	9. Norway	2.29
10. Norway	2.7	10. Norway	2.5	10. Philippines	1.95

3.2 million tons in 2002, and 2.9 million tons in 2004. Since then, an increasing trend has occurred, reaching 4 million tons in 2014 and 4.17 million tons in 2015.

China: From the late 1970s to the early 1980s, the annual national marine capture production stabilized at approximately 3 million tons (the overall national fisheries production was 4.5 million tons). Since implementation of the reform and opening up policy, the national fisheries production, including marine capture production, has continued to grow. In 1990, the marine capture production was 5.85 million tons. In 1991, this production exceeded 6 million tons (6.4 million tons), followed by 11.1 million tons in 1995, 13.1 million tons in 2009, and 14 million tons (14.13 million tons) in 2012 and reached 15.31 million tons in 2015.

The top 10 marine fishing states in the world in 2002, 2004, and 2015 are shown in Table 2.7, but the ranking was not fixed, with changes occurring at different times. The top 10 marine fishing states in 2002 were China, Peru, the USA, Indonesia, Japan, Chile, India, Russia, Thailand, and Norway. The top 10 in 2004 were China, Peru, the USA, Chile, Indonesia, Japan, India, Russia, Thailand, and Norway. The top 10 in 2015 were China, Indonesia, the USA, Peru, Russia, India, Japan, Vietnam, Norway, and the Philippines. In terms of ranking, China has always been at the top of marine capture production. Japan dropped from fifth place in 2002 to sixth place in 2004 and further dropped to seventh place in 2015. The USA and Norway have been relatively stable in the ranking, with the USA fixed in the third place and Norway fixed in ninth or tenth place. Among the top 10 marine fishing states, there have been 6 to 7 developing countries.

2.4.2 Main Marine Fishing Species

According to FAO statistics, the main global marine fishing species can be categorized into flatfish, cod, herring and anchovy, tuna, shrimp, cephalopod, and so on. Among them, shrimp includes prawn and other small shrimp, while

Cephalopoda includes squid, cuttlefish, and octopus. According to the statistical analysis of capture production from 2000 to 2015, flatfish production has fluctuated within an almost stable range of 0.86–1.05 million tons. However, the fluctuation in cod production is relatively large, fluctuating between 6.95 million and 9.39 million tons. Small- and medium-sized pelagic fish, such as sardines, are the most important fishing species. However, due to the unstable Peruvian anchovy production, its production has varied greatly from year to year, with the highest production of 24.75 million tons in 2000 and the lowest production of 15.59 million tons in 2014. Since 2000, the production levels of tuna and shrimp have increased steadily. In 2014, tuna production reached 7.48 million tons, which increased by 1.8 million tons compared to 5.68 million tons in 2000. Shrimp production also increased from 2.9 million tons in 2000 to 3.37 million tons in 2015. Cephalopod production also showed a steady growth trend, from 3.69 million tons in 2000 to 4.71 million tons in 2015 (Table 2.8).

According to FAO statistics, the main reasons for the interannual changes in production of the main global marine fishing species are overfishing, climate change, and improved fisheries management, which contribute to either reduced or increased fishery resources. As shown in Table 2.9, the top 10 marine fishing species in 2004 were anchovy, pollock, blue whiting, skipjack tuna, Atlantic herring, mackerel, Japanese anchovy, Chilean jack mackerel, hairtail, and yellowfin tuna, with annual productions ranging from 1.5–9.7 million tons. Compared to the top 10 list in 2002, capelin was the only species which was no longer listed among the top 10. The rankings of blue whiting and mackerel rose from the 8th and 9th to the 3rd and 6th positions, respectively, and the newly listed yellowfin tuna ranked 10th (Table 2.9). Additionally, in 2015, the top 10 global marine fishing species were anchovy, pollock, skipjack tuna, Atlantic herring, mackerel, blue whiting, yellowfin tuna, Japanese anchovy, Atlantic cod, and hairtail, with annual productions ranging from 1.27 to 4.31 million tons.

2.4.3 Development and Utilization of Fishery Resources in Different Fishing Areas

2.4.3.1 Northwest Pacific Ocean

The Northwest Pacific Ocean is FAO fishing area 61 (Fig. 2.15). The Bering Sea, Sea of Okhotsk, Sea of Japan, Yellow Sea, East China Sea, and South China Sea are included in this area. The main coastal states in this area are China, Russia, Japan, North Korea, South Korea, and Vietnam.

The Northwest Pacific Ocean is one of the world's most exploited fishing areas. This region has various fishery resources, especially pelagic fish, which fully reflects the natural conditions of topography, hydrology, and biology in this area. Cold and warm currents interact in this area, which not only affects the climatic conditions of the coastal areas but also creates favorable conditions for living resources in this

Table 2.8 Productions of main marine fishing species during 2000 to 2015 (unit: million tons)

Years	2000	2001	2002	2003	2004	2005	2006	2007
Flatfish	1.01	0.95	0.92	0.92	0.86	0.90	0.87	0.91
Cod	8.70	9.30	8.47	9.38	9.39	8.97	8.99	8.35
Anchovy sardine	24.75	20.44	22.14	18.66	23.02	22.28	19.18	20.14
Shrimp	2.90	2.78	2.73	3.22	3.24	3.13	3.24	3.19
Cephalopod	3.69	3.29	3.26	3.54	3.73	3.82	4.21	4.32
Tuna	5.68	5.64	5.98	6.14	6.35	6.53	6.55	6.63
Years	2008	2009	2010	2011	2012	2013	2014	2015
Flatfish	0.95	0.93	0.96	1.00	0.99	1.05	1.04	0.97
Cod	7.69	6.95	7.44	7.42	7.70	8.18	8.71	8.93
Herring sardine	20.39	20.18	17.27	21.17	17.57	17.60	15.59	16.70
Shrimp	3.05	3.08	3.01	3.21	3.26	3.23	3.30	3.37
Cephalopod	4.26	3.47	3.63	3.78	4.02	4.04	4.86	4.71
Tuna	6.52	6.61	6.64	6.59	7.09	7.22	7.48	7.39

Table 2.9 Top 10 marine fishing species in 2002, 2004, and 2015

	2002		2004		2015	
Rank	Species	Production (million tons)	Species	Production (million tons)	Species	Production (million tons)
1	Anchovy	9.7	Anchovy	10.7	Anchovy	4.31
2	Pollock	2.7	Pollock	2.7	Pollock	3.37
3	Skipjack tuna	2.0	Blue whiting	2.4	Skipjack tuna	2.82
4	Capelin	2.0	Skipjack tuna	2.1	Atlantic herring	1.51
5	Atlantic herring	1.9	Atlantic herring	2.0	Mackerel	1.49
6	Japanese anchovy	1.9	Mackerel	2.0	Blue whiting	1.41
7	Chilean jack mackerel	1.8	Japanese anchovy	1.8	Yellowfin tuna	1.36
8	Blue whiting	1.6	Chilean jack mackerel	1.8	Japanese anchovy	1.33
9	Mackerel	1.5	Hairtail	1.6	Atlantic cod	1.30
10	Hairtail	1.5	Yellowfin tuna	1.4	Hairtail	1.27

area. The warm Kuroshio Current and cold tidal currents intermingle in the sea area of Northeast Japan and develop into many eddy currents in the current boundary area, and the seawater is fully mixed. Research shows that in addition to counter-clockwise circulation in the Bering Sea and Okhotsk Sea, there is also a counter-clockwise circulation in the waters around the western Aleutian Islands to the southeast of Kamchatka. The favorable marine environmental conditions create excellent conditions for fishery resources and the formation of fishing grounds. The main fishing species in this area are sardine, anchovy, jack mackerel, mackerel, herring, Pacific saury, salmon and trout, skipjack, tuna, squid, pollock, flatfish, and whale, among others.

The Northwest Pacific Ocean has the highest production among the FAO fishing areas. Its total production fluctuated between 17 and 24 million tons from the 1980s to 1990s, and the annual fisheries production exceeded 20 million tons from 2010 to 2014. The three main fishing targets belong to pelagic fish, demersal fish, and crustaceans. The hairtail, pollock, and chub mackerel are the main capture species, as well as squid, cuttlefish, and octopus.

2.4.3.2   Northeast Pacific Ocean

The Northeast Pacific Ocean is FAO fishing area 67. The Northeast Pacific Ocean includes the eastern Bering Sea and the Gulf of Alaska (Fig. 2.15). There are many mountains, many islands, and some narrow coves along the coast of the Gulf of Alaska and some relatively shallow open waters in the eastern Bering Sea and the Chukchi Sea.

The main currents in the southern waters of the Aleutian Islands are the Alaska Current and the southern water system of the Alaska Current. The southern water system of the Alaska Current bifurcates at the nearshore of the USA at approximately 50°N, some flowing southward into the California Current and the remainder flowing northward into the Gulf of Alaska and then westward into the Alaska Current.

The main fishing species in this area are flounder, salmon, trout, and pollock in the Gulf of Alaska and the eastern Bering Sea. Squid and Pacific sardine are the main pelagic species. In addition, salmon from the Bering Sea and Alaska, king crab, and shrimp are also the main fishing species in the area.

The fisheries production in the Northeast Pacific Ocean is relatively low among all the fishing areas in the world. Its fishing production was below 2.5 million tons in the 1970s. The highest annual production of 3.3 million tons was recorded in the 1980s. Since then, production has fluctuated between 2.5 million and 3.2 million tons. From 2013 to 2014, the annual production exceeded 3 million tons. Demersal fish is the most important fish type in this fishing area, among which cod, hake, and haddock are the primary contributors to production. Among them, approximately 10% of fish species are overfished, 80% are fully fished, and 10% are underfished.

2.4.3.3   Western Central Pacific Ocean

The Western Central Pacific Ocean is FAO fishing area 71 (Fig. 2.15). The main fishing grounds are the continental shelf fishing grounds along the western coast and tuna fishing grounds around the central islands. Along the Western Central Pacific Ocean coast are China, Vietnam, Cambodia, Thailand, Malaysia, Singapore, East Timor, the Philippines, Papua New Guinea, Australia, Palau, Guam, Solomon Islands, Vanuatu, Micronesia, Fiji, Kiribati, Marshall, Nauru, New Caledonia, Tuvalu, and more.

This area is mainly affected by the North Equatorial Currents. The northern area is influenced by the Kuroshio Current, which has a stable flow. The surface current in the south is affected by the prevailing monsoon, and the flow direction varies under the influence of the monsoon. The North Equatorial Current flows westward to the north of 5°N, bifurcating into two branches in the Philippines, with one flowing northward and the other southward. The northward branch flows north along the eastern coast of the Philippine islands and then turns northeast at the eastern coast of Taiwan, developing into the Kuroshio Current. However, the southward branch flows into Southeast Asia during a certain season. In February, the northeast monsoon prevails to the north of the equator. The northern equatorial water flows into Southeast Asia along the southern side of the Philippines. The current in the South China Sea flows southward along the Asian continent, and a large amount flows into the Java Sea and then to the Indian Ocean through the Banda Sea. A small branch flows into the Indian Ocean through the Strait of Malacca. In August, the southern equatorial flow goes into Southeast Asia with a strong flow. In the southern seas, the surface laminar circulation usually goes into the Java Sea through the Banda Sea, and a large amount of Pacific water enters the Indian Ocean through the Timor Sea. At the same time, the currents of the South China Sea flow northward along the continental shelf.

This region is one of the areas with developed fisheries, various small fishing vessels conducting fishing, varieties of fishing gear used, and varieties of fishing species targeted. The Western Central Pacific Ocean is also a fishing area with high potential production. The production has steadily increased since the 1970s. The production rose from less than 4 million tons in 1970 to more than 12.5 million tons in 2014. In this fishing area, pelagic fish and demersal fish are the main fishing targets, among which skipjack tuna is the main target.

2.4.3.4 Eastern Central Pacific Ocean

The Eastern Central Pacific Ocean is FAO fishing area 77 (Fig. 2.15). Coastal states in this area include the USA, Mexico, Guatemala, El Salvador, Ecuador, Nicaragua, Costa Rica, Panama, and Colombia, among others. The long coastline (approximately 9000 km, excluding the Gulf of California) is largely mountainous with narrow continental shelves. There are some islands in southern California and Panama, few islands and shoals in the open water, and some isolated islands or archipelagos, such as Clifton Island and the Galapagos Islands. Moreover, there are only narrow island shelves around the islands. These islands or archipelagos cause local hydrological changes that attract tuna and other pelagic fish to gather there, greatly promoting the fisheries in this area.

Two surface currents flow through this area. One is the California Current in the north, and the other is the Peruvian Current in the south. In addition, the subsurface equatorial countercurrent is also an important current. The California Current flows southward along the US coast and is affected by the prevailing north and northwest winds, generating strong upwelling and peaking in summer. In winter, the north wind weakens or blows the south wind, and there is a countercurrent flowing along the coast, so the hydrological structure near the shore is more complicated. There are semi-permanent eddies around the islands of southern California. A branch of the California Current flows along the coast of Central America to the low latitudes of the Eastern Pacific Ocean, turning west at approximately 10°N and merging with the North Equatorial Current. Near the coast, the equatorial countercurrent flows mostly along the Central American coast to the north, forming the Costa Rican Current. Eventually, it merges with the Equatorial Currents, creating a counterclockwise vortex in the open waters of Costa Rica, giving rise to the Costa Rica Dome with its center between 7°N and 9°N and 87°W to 90°W, where the lower seawater rises.

Pelagic fish, such as sardine, anchovy, jack mackerel, and tuna, are the main fishing species in this area. The total production is not high, at less than 2 million tons overall. The production in the Eastern Central Pacific has fluctuated between 1.2 and 2 million tons since 1980, remaining at 1.8–2 million tons from 2011 to 2014. The production in 2010 was approximately 2 million tons.

2.4.3.5 Southwest Pacific Ocean

The Southwest Pacific Ocean is FAO fishing area 81 (Fig. 2.15). This fishing area has many islands, such as New Zealand and Easter Island. This area is vast, and almost all areas are deep water, with the two coastal states of Australia and New Zealand. Its continental shelf is mainly distributed around New Zealand and the eastern and southern coasts of Australia (including the southwestern coast of New Guinea). The main fishing areas are around Australia and New Zealand.

The hydrological situation in the South Pacific (especially areas far from South America and the Australian coast) is unclear. The main currents in this area are the South Equatorial Current and the Drift Current in the northern area, as well as the West Wind Drift in the southern part. In the Tasman Sea, the East Australian Current flows southward along the Australian coast, which weakens and spreads around south of Sydney. The currents around New Zealand are complex and variable.

The production in the Southwest Pacific is not high, and, actually, it is the lowest currently. The maximum is less than 900,000 tons, and the main fishing targets are demersal fish. Its production peaked in 1998, reaching 857,000 tons. Since 2000, there has been a slight downward trend in capture production, with 500,000–600,000 tons from 2008 to 2014. Species such as the Wellington flying squid are the main fishing species.

2.4.3.6 Southeast Pacific Ocean

The Southeast Pacific Ocean is FAO fishing area 87 (Fig. 2.15). Coastal states include Colombia, Ecuador, Peru, and Chile. Extensive upwellings occur in the sea area. The main fishing ground is the area around the continental shelf along the western coast of South America.

The Peruvian Current is the main ocean current in the area, and it forms a wide upwelling when it goes northward, creating conditions for fishing ground formation. The main fishing species in this area is anchovy, which is distributed along the entire coast of Peru and the northernmost part of Chile, followed by Chilean jack mackerel, South American sardine, mollusk, tuna, and other species.

The Southeast Pacific Ocean is the most significant fishing area in the world with a recorded annual production at approximately 20 million tons. Marine capture production peaked at 20.31 million tons in 1994. Pelagic fish are the main fishing target in this area. Therefore, capture production varies from year to year. From 2012 to 2014, the production varied from 6 to 8 million tons.

2.4.3.7 Northwest Atlantic Ocean

The Northwest Atlantic Ocean is FAO fishing area 21 (Fig. 2.15). This area is mainly located on the west coast of Greenland, with Newfoundland as the center, and the northeastern coast of North America. The main part of this area falls under management of the International North Atlantic Fisheries Commission (ICNAF).

The main currents in this area are the high-temperature and high-salinity Gulf Current and Labrador Cold Current. They converge at the great shoals in the south of Newfoundland and form the world-famous Newfoundland fishing ground. Considering the fishing gear, bottom trawling and longline fisheries are the two primary fisheries in this area. Considering the fishing species, the two primary fisheries are the menhaden and oyster fisheries, and menhaden is mainly processed as raw material for fishmeal and fish oil. The main catches are cod, haddock, perch, hake, herring, and other demersal fish and pelagic fish (such as salmon).

According to FAO production statistics, production in this area was stable at above 4 million tons in the early 1970s, followed by a continuous decline. In the 1980s, production was stable from 2.5 to 3 million tons and was approximately 2 million tons from 2000 to 2010. From 2011 to 2014, production was less than 2 million tons, of which the production of pelagic fish, crustaceans, and mollusks accounted for the majority. An estimated 77% of species are fully fished, 17% are overfished, and 6% are underfished.

2.4.3.8 Northeast Atlantic Ocean

The Northeast Atlantic Ocean is FAO fishing area 27 (Fig. 2.15), including countries such as Portugal, Spain, France, Belgium, the Netherlands, Germany, Denmark, Poland, Finland, Sweden, Norway, Russia, the UK, Iceland, Greenland, and Novaya Zemlya. This area is also the statistical area of ICES (International Council for the Exploration of the Sea). The North Sea fishing grounds, Icelandic fishing grounds, northern fishing grounds of Norway, southeastern fishing grounds of the Barents Sea, and the continental shelf fishing grounds stretching from Bear Island to Spitsbergen are the main fishing grounds in this area.

This area is dominated by the warm North Atlantic Current and its tributaries. At the southern coast of Iceland, the Irminger Current (warm current) flows westward, while the Eastern Icelandic Current (cold current) flows near the northern and eastern coast. The North Atlantic Current flows northward along the western coast of Norway after passing through Faro Island and then divides into two branches, one continuously flowing northward to the western coast of Spitsbergen and the other flowing northeast along the Norwegian northern coast into the Barents Sea. These two currents warm up the sea in the western and southern parts of the Barents Sea and promote the production there.

This area has some fisheries with the longest history in the world. The North Sea fishing grounds are one of the three most famous fishing grounds in the world and are the cradle of modern trawling fisheries. The main fishing grounds for trawling are the areas around Dogger Bank and Great Fisher Bank. The salmon fishery conducted in inshore Norway and Iceland as well as the North Sea fishing grounds is the most important and longest-established fishery in the world. The main catches are cod, haddock, hake, Norway pout, pollock, Clupeidae, and mackerel. Among these fish, the herring production is the highest, with an annual production of 2–3 million tons, followed by cod, with an annual production over 2 million tons.

Production in the Northeast Atlantic significantly declined after 1975, but recovered in the 1990s. The production in 2010 was 8.7 million tons. According to FAO statistics, capture production from 2011 to 2014 stabilized between 8 and 8.5 million tons. The main fishing species are demersal fish and pelagic fish. The FAO estimates that, overall, 62% of species are fully fished, 31% are overfished, and the remaining 7% are underfished.

2.4.3.9   Western Central Atlantic Ocean

The Western Central Atlantic Ocean is FAO fishing area 31 (Fig. 2.15). The main countries in this area are the USA, Mexico, Guatemala, Honduras, Nicaragua, Costa Rica, Panama, Colombia, Venezuela, Guyana, Suriname, and some island states, such as Cuba, Jamaica, Haiti, and Dominica in the Caribbean. The main fishing grounds in this area are located in the Gulf of Mexico and the Caribbean Sea.

The main current of this area is a tributary of the Equatorial Current, which goes westward along the coast of South America to the Caribbean Sea together with the Equatorial Current, forming the Caribbean Current. Then, this current flows strongly westward and forms an upward flow near the coast of Venezuela and Colombia under the influence of the wind. The Caribbean Current flows out of the Caribbean Sea and forms a clockwise circulation in the eastern Gulf of Mexico after passing through the Yucatan Channel. This current becomes the strong Florida Current after leaving the Gulf of Mexico, constituting the beginning of the Gulf Stream system, and flows northward to the eastern coast of the USA.

The main catches in this area are shrimp and pelagic fish. According to FAO statistics, the total production in the Western central Atlantic reached a maximum of 2.5 million tons in 1984 and then gradually declined. The production from 1995 to 2005 stabilized between 1.7 and 1.83 million tons. The production was further reduced to approximately 1.2 million tons in 2005 and from 1.2 to 1.4 million tons between 2011 and 2014.

2.4.3.10 Eastern Central Atlantic Ocean

The Eastern Central Atlantic Ocean is FAO fishing area 37 (Fig. 2.15). The Mediterranean Sea and the Black Sea are in this area. The main countries in the region are Angola, Congo, Gabon, Equatorial Guinea, Cameroon, Nigeria, Benin, Togo, Ghana, Ivory Coast, Liberia, Sierra Leone, Guinea, Guinea-Bissau, Senegal, Mauritania, Western Sahara, Morocco, and countries along the coast of Mediterranean Sea.

The main surface currents in this area are the Canary Current flowing from north to south and the Benguela Current flowing from south to north. Arriving near the equator, the two currents merge into the North Equatorial Current and the South Equatorial Current in the west. There is an equatorial countercurrent between the two main streams, and its tributary flow, the Guinea Current, goes westward to the Gulf of Guinea. Additionally, in the inshore Ivory Coast, there is a countercurrent of the Guinea Current along the western coast. The Canary Current (cold current) going southward along the northern water area of West Africa and the equatorial countercurrent (warm current) going northward along the southern coast of West Africa merge at the northern water area of West Africa, forming a seasonal upwelling. Due to the extensive range of the continental shelf, great fishing grounds have formed there.

Pelagic fish are the main fishing targets in the area, mainly including sardine, jack mackerel, mackerel, longfin tuna, yellowfin tuna, and tuna. Large-sized demersal fish, such as bream, and cephalopods, such as squid, cuttlefish, and octopus, are also major fishing species.

The Eastern Central Atlantic Ocean has also been an important fishing area for deep-sea fishing states since the 1970s, but the total production of this area has fluctuated, although it has shown an increasing trend overall. The capture productions from 2010 to 2014 stabilized between 4 and 4.5 million tons. According to the assessment, sardines (in the area from Cape Bojadour south to Senegal) are still considered underfished. In contrast, most pelagic species, such as the small sardine in Northwest Africa and the Gulf of Guinea, are considered fully fished or overfished. Most of the demersal fish have a status between fully fished and overfished in most areas, while the whitespotted grouper in Senegal and Mauritania is still in a serious situation. Some deep-water shrimp species are fully fished, while other shrimp species have a status between fully fished and overfished. Octopus and cuttlefish are still overfished. Overall, an estimated 43% species are fully fished in the Eastern Central Atlantic Ocean, 10.5% are overfished, and the remaining 4% are underfished. Therefore, this area calls for scientific management for fisheries resource conservation.

2.4.3.11   Southwest Atlantic Ocean

The Southwest Atlantic Ocean is FAO fishing area 41 (Fig. 2.15), including countries such as Brazil, Uruguay, and Argentina. The main fishing grounds in this area
are located on the continental shelf of the eastern coast of South America. The
continental shelf of this sea area is affected by two major currents, the northern being
the warm Brazilian Current and the southern being the cold Falkland Current. The
two currents merge here and form a wide interchange area.

Fisheries in this area are mostly conducted by local residents. Small fishing boats
and bamboo rafts are widely used in fisheries in the northern and central coasts of
Brazil, while in the southern coasts of Brazil and Patagonia, large bottom trawls are
widely used. In Uruguay and Argentina, various types of trawls are widely used in
fisheries production. Argentine hake, Argentine short-finned squid, sardine, and
mackerel are the main catches in this area.

The Southwest Atlantic Ocean is also an important fishing area for deep-sea
fishing states. Its production has continued to grow since the 1970s, and this area
reached a peak production of more than 2.6 million tons in 1997. The production of
Illex argentinus has led to a dynamic change in the annual production in this area. Its
production has been stable at 1.7–2.5 million tons since 2010. It is estimated that the
main fishing species, such as Argentine hake and Brazilian sardines, are still
overfished, although the latter is showing signs of recovery. Among the monitored
fish species in this region, 50% are fully fished, 41% overfished, and the remaining
9% underfished.

2.4.3.12   Southeast Atlantic Ocean

The Southeast Atlantic Ocean is FAO fishing area 47 (Fig. 2.15). The main fishing
grounds are on the continental shelf area along the western coast of South Africa.
Angola, Namibia, and South Africa are the coastal states of this area. The main
current of this sea area is the Benguela Current, which flows northward from 3°S to
15°S along the western coast of Africa and then forms the South Equatorial Current
in the west. The Benguela Current flows northward along the western coast of
southern Africa, generating upwelling under the influence of offshore winds. The
range of its influence varies from season to season. However, the main current in the
south is the West Wind Drift.

The Southeast Atlantic Ocean is also an important fishing area for deep-sea
fishing states. The main fishing targets in this area are pelagic and demersal fish.
Production in the Southeast Atlantic has shown an overall downward trend since the
early 1970s. The production in this area was 3.3 million tons in the late 1970s but
dropped to only 1.2 million tons in 2009 and was stable at approximately 1.5 million tons from 2012 to 2014. According to assessment, hake is still in a state of being
fully fished and overfished. The deep-water hake in South African waters and the
South African hake in Namibian waters show some signs of recovery. South African

sardine varies widely and has a large biomass. Under adverse environmental conditions, South African sardine resources have fallen sharply and are considered fully fished or even overfished. South African anchovy resources have been maintaining recovery and are considered fully fished. The situation with jack mackerel is serious, especially in Namibian and Angolan waters, where the resources are overfished.

2.4.3.13 Mediterranean Sea and Black Sea

The Mediterranean Sea (FAO fishing area 37) is a large, almost closed water body that separates Europe from Africa and Asia. The Mediterranean Sea is divided by the Tunisian Strait into two parts, the Eastern Mediterranean and the Western Mediterranean (Fig. 2.15). Atlantic waters enter the Mediterranean Sea through the Strait of Gibraltar and flow mainly along the African coast to the Eastern Mediterranean Sea. The lower salinity water of the Black Sea enters the Mediterranean Sea through the surface currents. The Nile is the main source of freshwater in the Mediterranean area, affecting the hydrology, productivity, and fisheries in the Eastern Mediterranean Sea. The construction of the Aswan Dam has changed the ecological environment and directly affected fisheries development. The Suez Canal brings high-temperature surface water from the Red Sea into the Mediterranean Sea, while the cold bottom water enters the Red Sea from the Mediterranean Sea.

In general, the Mediterranean Sea lacks fishery resources, which are rich in species but poor in quantity. Large-scale fisheries are mainly conducted in the Black Sea. Small-scale fisheries in the Mediterranean are developed, and regional resources have been fully fished or overfished, with demersal fish resources being the most fully utilized. The pelagic fisheries production accounts for approximately half of the total production. The main catches are sardine, Sprattus, skipjack tuna, and tuna. Additionally, hake is the main catch for demersal fisheries.

According to FAO statistics, in the mid-1980s, fishing production peaked at approximately 1.9 million tons, and it stabilized at 1.4–1.7 million tons from 1996 to 2008. However, the production dropped at 1.1–1.3 million tons during 2012–2014. According to analysis, all European hake and goatfish resources have been overexploited and so have most species of sole and bream. The main species of small pelagic fish (sardine and anchovy) have been assessed as fully fished or overfished. In the Black Sea, small pelagic fish (Sprattus and anchovy in particular) have recovered to some extent from the great recession caused by unfavorable marine conditions in the 1990s but are still considered fully fished or overfished, while most other species may have a status between fully fished and overfished. Overall, 33% of the species in the Mediterranean Sea and the Black Sea are assessed as fully fished, 50% as overfished, and the remaining 17% as underfished.

2.4.3.14   Eastern Indian Ocean

The Eastern Indian Ocean is FAO fishing area 57 (Fig. 2.15). This area mainly includes regions such as eastern India, western Indonesia, Bangladesh, Vietnam, Thailand, Myanmar, and Malaysia, and it is rich in such fishery resources as herring, sardine, milkfish, and shrimp, among others.

The main fishing grounds in the Eastern Indian Ocean are continental shelf fishing grounds and tuna fishing grounds. Its coastal states include India, Bangladesh, Myanmar, Thailand, Indonesia, and Australia. The catches in this area are mainly from the coastal states. There are few fishing vessels from the deep-sea fishing states operating in this fishing area. At present, countries and regions that operate in this fishing area are China, Japan, Chinese Taipei, France, South Korea, and Spain, and their main fishing target is tuna.

According to statistics, fishing production in the Eastern Indian Ocean has maintained a high growth rate, from more than 1 million tons in 1970 to nearly 8 million tons in 2014, the highest growth rate in all fishing areas. The production is mainly from pelagic fish and demersal fish, while some production (approximately 42% of the overall production in this area) is categorized as from "undetermined marine fish species." Increased production may be the result of fishing expansion or the exploration of new species in new areas. Total production in the Bay of Bengal and the Andaman Sea has grown steadily, with no signs of resource depletion.

2.4.3.15   Western Indian Ocean

The Western Indian Ocean is FAO fishing area 51 (Fig. 2.15). The important coastal states are India, Sri Lanka, Pakistan, Iran, Oman, Yemen, Somalia, Kenya, Tanzania, Mozambique, South Africa, Maldives, and Madagascar, among others. The main catches in this area are sardine, drum fish, skipjack tuna, yellowfin tuna, Bombay duck, Spanish mackerel, hairtail, and shrimp.

The main fishing grounds in the Western Indian Ocean are continental shelf fishing grounds and tuna fishing grounds. The catches in this area are mainly from the coastal states, embodying 90.6% of the total catches. The countries and regions that are currently engaged in fishing in this area are Japan, France, Spain, South Korea, and Chinese Taipei. Tuna and demersal fish are the main catches, accounting for less than 10% of the total.

According to FAO statistics, production in the Western Indian Ocean continued to grow from less than 1.5 million tons in 1970 to approximately 4.5 million tons in 2006. Since then, it has declined slightly, and the production reported in 2010 was 4.3 million tons. The production further increased from 2012 to 2014, reaching more than 4.6 million tons in 2014. An assessment showed that Spanish mackerel in the Red Sea, the Arabian Sea, the Gulf of Oman, the Persian Gulf, and the coasts of Pakistan and India is overfished. The Southwest Indian Ocean Fisheries Commission conducted an assessment of 140 fish species. Overall, approximately 65% of fish species are fully fished, 29% overfished, and 6% underfished.

2.4.3.16 Southern Ocean

FAO fishing areas 48, 58, and 88 are in the Southern Ocean. They are located south of 60°S in the Pacific and West Atlantic, bounded by 45°S in the East Atlantic and 55°S in the Indian Ocean (Fig. 2.15). This region is adjacent to FAO fishing areas 81, 87, 41, 47, 51, and 57 and is in the Southern Ocean area. The Southern Ocean is connected to three other oceans. In the winter, half of the Southern Ocean is covered by ice. The Southern Ocean can be divided into the Atlantic Antarctic, the Pacific Antarctic, and the Indian Antarctic Ocean. Krill resources are rich in the Southern Ocean, while the fish species are limited, with only a few economic species, such as Antarctic fish and icefish.

Analysis and research have been carried out on the main surface currents of the Southern Ocean. There are upflows at approximately 65°S, where a low-pressure convergence is located, and there is also upwelling near the Antarctic continent. The Southern Ocean is a deep-sea system with significant circulation, and upwelling brings rich nutrients to the surface. Therefore, the production is very high in summer, while it is significantly reduced in winter.

According to a survey, there are approximately 60 pelagic fish species in the Southern Ocean (including sub-Antarctic waters) and approximately 90 demersal fish species, but the quantity of these fish species is still unclear. In waters with rich nutrition (convergence zones in particular) in the Pacific Antarctic, based on an estimate of the average dry weight of the lantern fish at 0.5 g/m^2, the production of mid-level trawling from the Russian fishing fleet is 5–10 t per 2 h.

The main living resource in the Southern Ocean area is krill. Scientists have completely different estimates of krill resources. Based on the quantity of krill eaten by a whale, it is estimated that the total biomass of krill resources is between 150 million and 5 billion tons. Based on the primary productivity data of the Southern Ocean, it is estimated that the total biomass is 500 million tons and that the annual catch is 100–200 million tons. It is also believed that the total krill biomass is 210–290 million tons, the amount consumed by predators such as whales is 130–140 million tons/year, and the fishable krill is less than 40–50% of the total biomass.

2.5 Global Aquaculture

2.5.1 Main Aquaculture Countries

In the 1980s, aquaculture began to attract global attention, and since then, it has quickly developed. In 1985, the total global aquaculture production (including aquatic plant production, the same below) exceeded 10 million tons by reaching 11.06 million tons. From then on, it has continued to rise, reaching 90 million tons in 2012 and 106 million tons in 2015. The rapid development of the global aquaculture industry not only compensates for the fluctuation or decline of capture production, caused by the decline of the main economic fishery resources due to overfishing, but also plays an important role in improving the structure of human diet consumption. The continued development of the global aquaculture industry strongly correlates with China's vigorous development of inland and marine aquaculture. China has effectively promoted aquaculture development in developing countries with abundant water resources. The aquaculture industry has also been prioritized in developed countries that have long been engaged in marine capture. For example, Norway has vigorously developed Atlantic salmon aquaculture, which has brought about significant economic, social, and ecological benefits and has promoted marine aquaculture development in other countries.

The top 10 aquaculture countries in the world in 2000, 2005, 2010, and 2015 are shown in Table 2.10. Top ranking countries in terms of aquaculture production in 2000 were China, India, Japan, Indonesia, Thailand, Bangladesh, Vietnam, Norway, the USA, and Chile. In 2005, the rankings of the top 10 aquaculture countries in the world changed. Most notably, Japan fell from third to sixth, and Vietnam rose from seventh to fourth. The top 10 aquaculture countries in the world in 2010 were basically of the same as those in 2005. Remarkably, Bangladesh has risen from 7th to 5th. The top 10 aquaculture countries in the world in 2015 were basically the same as those in 2010.

There are some highlights in Table 2.10. First, aquaculture production in China is very extensive, and China is the primary contributor to the production of the global aquaculture industry. Second, the top 10 aquaculture countries in the world remained basically unchanged from 2005 to 2015, but there have been significant changes in

Table 2.10 Top 10 aquaculture countries in the world from 2000 to 2015 (unit: 10,000 tons)

	2000	2005	2010	2015
Bangladesh	65.71[6]	88.21[7]	130.85[5]	206.04[5]
Chile	42.51[10]	73.94[8]	71.32[9]	105.77[8]
China	2846.02[1]	3761.49[1]	4782.96[1]	6153.64[1]
India	194.25[2]	297.31[2]	379.00[3]	523.80[3]
Indonesia	99.37[4]	212.41[3]	627.79[2]	1564.93[2]
Japan	129.17[3]	125.41[6]	115.11[7]	110.32[7]
Norway	49.13[8]	66.19[9]	101.98[8]	138.09[6]
Thailand	73.82[5]	130.42[5]	128.61[6]	89.71[9]
USA	45.68[9]	51.39[10]	49.67[10]	42.60[10]
Vietnam	51.35[7]	145.23[4]	270.13[4]	345.02[4]
Total production (percentage of global aquaculture production)	3597.02 (86.21%)	4952.01 (85.64%)	6657.44 (85.33%)	9279.92 (87.54%)
Global aquaculture production	4172.46	5782.02	7802.00	10600.42

Note: The rankings are indicated in [ ]

the ranking and production. Notably, in Japan and the USA, where fisheries are highly developed, the aquaculture productions remained basically unchanged, stabilizing between 1.1 and 1.3 million tons and 400,000–500,000 tons, respectively. However, aquaculture production in Norway significantly increased, from less than 500,000 tons in 2000 to more than 1.38 million tons in 2015. Third, the total production from the top 10 aquaculture countries accounts for 85–88% of the world total, while the total production from other countries is less than 12%. Fourth, 7 of the top 10 aquaculture countries are developing countries. In fact, aquaculture is mainly conducted in developing countries.

2.5.2 Main Aquaculture Species

With the development of artificial breeding and culture techniques, the number of aquaculture species has gradually increased, shifting from common species to well-known, special, and desirable species. According to FAO reports, there were 34 families and 72 species included in aquaculture in 1950, while in 2004, aquaculture had expanded to 115 families and 336 species. The aquaculture species in different areas in 2004 are shown in Table 2.11. This table shows that the Asian and Pacific areas have the largest number of aquaculture species.

The FAO statistics on the volume and value of aquaculture species from 2000 to 2015 are shown from Tables 2.12, 2.13, and 2.14. Species sorted by production are

Table 2.11 Aquaculture species in different areas in 2004

	Family	Species
Global	245	336
North America	22	38
Middle and Eastern Europe	21	51
Latin America and the Caribbean Sea	36	71
Western Europe	36	83
Sub-Saharan Africa	26	46
Asia and the Pacific Ocean	86	204
Middle East and North Africa	21	36

Table 2.12 Production of aquaculture species from 2000 to 2015 (unit: million tons)

	2000	2005	2010	2015
Aquatic plants	9.31	13.50	18.99	29.36
Crustaceans	1.69	3.78	5.59	7.35
Migratory fish	2.25	2.87	3.61	4.98
Freshwater fish	17.59	23.68	33.00	44.05
Marine fish	0.98	1.44	1.88	2.88
Other aquatic animals	0.16	0.45	0.88	0.95
Mollusks	9.76	12.11	14.07	16.43

Table 2.13 Production value of aquaculture species from 2000 to 2015 (unit: million USD)

	2000	2005	2010	2015
Aquatic plants	2909.38	3887.22	5642.44	4846.89
Crustaceans	9425.86	14947.18	26825.63	38519.90
Migratory fish	6470.99	9497.75	15892.80	20108.61
Freshwater fish	18403.15	24476.61	51070.36	67459.20
Marine fish	4223.27	5254.78	8350.29	10238.09
Other aquatic animals	929.89	1905.13	3314.13	3948.30
Mollusks	8712.16	10192.17	14325.52	17853.60

Table 2.14 Production value of aquaculture species per ton from 2000 to 2015 (unit: million USD/tons)

	2000	2005	2010	2015
Aquatic plants	312.63	287.88	297.09	165.07
Crustaceans	5573.32	3956.32	4802.64	5239.84
Migratory fish	2874.84	3314.17	4398.66	4035.92
Freshwater fish	1046.49	1033.81	1547.58	1531.56
Marine fish	4323.05	3653.14	4438.01	3556.12
Other aquatic animals	5947.70	4275.74	3760.16	4155.16
Mollusks	892.86	841.41	1018.45	1086.51

listed as follows: freshwater fish, aquatic plants, mollusks, crustaceans, migratory fish between fresh and seawater, marine fish, and other aquatic animals, with no interannual variations. Species sorted by production value (referencing 2015 as an example) are listed as follows: freshwater fish, crustaceans, migratory fish between fresh and seawater, mollusks, marine fish, aquatic plants, and other aquatic animals. Species sorted by production value per ton are listed as follows: crustaceans, other aquatic animals, migratory fish between fresh and seawater, marine fish, freshwater fish, mollusks, and aquatic plants. As shown in Table 2.14 (referencing 2015 as an example), the production value per ton of crustaceans, migratory fish, and marine fish is equivalent to 3.42 times, 2.64 times, and 2.32 times that of freshwater fish, respectively.

According to FAO statistics, from 1970 to 2015, crustaceans (20.71%) had the highest annual growth rate of production among aquaculture species in the world, followed by marine fish and freshwater fish, with annual growth rates of 12.22% and 11.13%, respectively. Additionally, the average annual growth rates of aquatic plants and migratory fish were 10.27% and 8.82%. The average annual growth rate of production of different aquaculture species varies from year to year, as shown in Table 2.15.

The aquaculture species with annual production of more than 2 million tons in 2015 are listed as follows: Atlantic salmon, bighead carp, common carp, grass carp, silver carp, and *Penaeus vannamei*, with productions of 2.3816 million tons,

Table 2.15 Annual average growth rate of production of aquaculture species in the world from 1970 to 2015 (%)

	1970–1980	1980–1990	1990–2000	2000–2010	1970–2015
Aquatic plants	10.67	3.61	9.47	7.39	10.27
Crustaceans	23.94	24.17	8.40	12.69	20.71
Migratory fish	6.52	9.53	6.43	4.85	8.82
Freshwater fish	5.88	13.91	9.43	6.50	11.13
Marine fish	13.87	5.82	11.52	6.77	12.22
Mollusks	5.57	6.99	10.46	3.72	8.12
Total production value	7.63	8.88	9.48	6.38	10.26

Table 2.16 Aquaculture species with annual productions over 1 million tons from 1970 to 2015 (unit: 10, 000 tons)

	1970	1980	1990	2000	2010	2015
Atlantic salmon	0.03	0.53	22.56	89.58	143.71	238.16
Bighead carp	12.49	19.86	67.80	142.82	258.70	340.29
Common carp	17.32	24.65	113.43	241.04	342.06	432.81
Grass carp	9.26	15.46	105.42	297.65	436.23	582.29
Silver carp	26.72	41.76	152.05	303.47	409.97	512.55
Penaeus vannamei	0.01	0.84	9.17	15.45	268.82	387.98

3.4029 million tons, 4.3281 million tons, 5.8229 million tons, 5.1255 million tons, and 3.8798 million tons, respectively. Among them, *Penaeus vannamei*, Atlantic salmon, grass carp, and bighead carp were the species with the fastest growth in aquaculture production between 1990 and 2015, which increased by 42.31 times, 10.56 times, 5.52 times, and 5.02 times, respectively (Table 2.16).

2.5.3 Current Status of Global Aquaculture Development

2.5.3.1 Overview

In 2014, aquaculture production amounted to 73.8 million tons, consisting of 49.8 million tons of fish, 16.1 million tons of mollusks, 6.9 million tons of crustaceans, and 7.3 million tons of other aquatic animals including salmon. Almost all aquaculture fish are destined for human consumption, although by-products may be used for non-food purposes. Fish from global aquaculture accounted for 44.1% of total production (including those for non-food uses) from both capture and aquaculture in 2014, with 42.1% in 2012 and 31.1% in 2004. All continents have shown a general trend of an increasing share of aquaculture production in total global fish production, although in Oceania, this share declined over the last 3 years.

At the national level, in 35 countries whose combined population is 3.3 billion, that is, 45% of the global population, aquaculture production outpaced capture production in 2014. Among these 35 countries, 5 countries are the most important contributors to aquaculture production, and they are China, India, Vietnam, Bangladesh, and Egypt. The other 30 countries also have well-developed aquaculture sectors, e.g., Greece, the Czech Republic, and Hungary in Europe and the Lao People's Democratic Republic and Nepal in Asia.

In addition to fish production, aquaculture produces considerable quantities of aquatic plants. Global aquaculture production of fish and aquatic plants reached 101.1 million tons of live weight in 2014, with farmed aquatic plants contributing 27.3 million tons. Thus, farmed fish production constitutes three-quarters of total aquaculture production by volume and farmed aquatic plants one-quarter, but the latter's share of total aquaculture value is disproportionately low (less than 5%).

The global production volume of farmed fish and aquatic plants surpassed that of capture production in 2013. In terms of food supply, aquaculture provided more fish than capture fisheries for the first time in 2014.

2.5.3.2 Fed and Non-fed Aquaculture Production

Feed is widely regarded as a major constraint to the growth of aquaculture production in many developing countries. However, by volume, half of the global aquaculture production in 2014, including seaweed and microalgae (27%) and filter-feeding animal species (22.5%), was from non-fed aquaculture.

The production of non-fed aquaculture animals in 2014 was 22.7 million tons, representing 30.8% of the global production of all farmed fish species. The most important non-fed aquaculture animal species include (1) silver carp and bighead carp for inland aquaculture, (2) bivalve mollusks (clam, oyster, mussel, among others), and (3) other filter-feeding animals (such as sea squirt) in marine and coastal areas.

Europe produced 632,000 tons of bivalves in 2014, and its major producers were Spain (223,000 tons), France (155,000 tons), and Italy (111,000 tons). Bivalve culture in China in 2014 provided approximately 12 million tons of product, 5 times that produced in the rest of the world. Other major Asian bivalve producers are Japan (377,000 tons), South Korea (347,000 tons), and Thailand (210,000 tons).

The production growth of fed-aquaculture species has been faster than non-fed species, although production of non-fed species can be more advantageous in terms of food security and environmental friendliness. The usually lower cost, non-fed aquaculture is largely undeveloped in Africa and Latin America and may offer development potential through species diversification to improve national food security and nutrition in these regions. Of the 8.2 million tons of global production

Table 2.17 Aquaculture production by region and percentage of global total production from 1995 to 2014

	Unit	1995	2000	2005	2010	2012	2014
Africa	Production (thousand tons)	110.2	399.6	646.2	1285.6	1484.3	1710.9
	Percentage (%)	0.45	1.23	1.46	2.18	2.23	2.32
Americas	Production (thousand tons)	919.6	1423.4	2176.9	2514.2	2988.4	3351.6
	Percentage (%)	3.77	4.39	4.91	4.26	4.50	4.54
Asia	Production (million tons)	21677.5	28422.5	39188.2	52439.2	58954.5	65601.9
	Percentage (%)	88.91	87.68	88.47	88.92	88.70	88.91
Europe	Production (thousand tons)	1580.9	2050.7	2134.9	2544.2	2852.3	2930.1
	Percentage (%)	6.48	6.33	4.82	4.31	4.29	3.97
Oceania	Production (thousand tons)	94.2	121.5	151.5	189.6	186.0	189.2
	Percentage (%)	0.39	0.37	0.34	0.32	0.28	0.26
Global	Production (thousand tons)	24382.4	32417.7	44297.7	58972.8	66465.5	73783.7

of filter-feeding fish from inland aquaculture in 2014, China produced 7.4 million tons, and the rest was produced in more than 40 other countries.

2.5.3.3 Distribution of Aquaculture Production and Aquaculture Production Per Capita

There has been no change in the overall pattern of uneven production distribution among regions and among countries in the same region (Table 2.17). Asia has accounted for approximately 89% of fish products for human consumption from aquaculture for the past two decades. Africa and the Americas have improved their respective shares in the global total production, while those of Europe and Oceania have dropped slightly.

The aquaculture development rate has outpaced the population growth rate, with aquaculture production per capita increasing in most regions over the past 30 years. As a whole, Asia outpaces other continents in fish production for human consumption, but there are vast regional differences in Asia.

In 2014, there were 25 countries with aquaculture production exceeding 200,000 tons, which produced 96.3% of global farmed fish and 99.3% of farmed aquatic plants. The farmed species and their importance in total national production vary widely among the main aquaculture countries. To date, China is still the primary producer, although its share of global farmed fish has fallen from 65% to less than 62% over the past 20 years.

2.6 Global Aquatic Products Processing and Utilization

2.6.1 *Status of Global Aquatic Products Processing and Utilization*

Capture and aquaculture products are very heterogeneous in terms of species and product forms. The products from one species can be prepared in many different ways and processed into different foods, a characteristic that makes fish a very versatile raw material for food. However, fish is also highly perishable and can spoil more rapidly than almost any other food, and if perished or spoiled, fish is inedible or even dangerous to health due to microbial growth, chemical change, and broken down endogenous enzymes. Therefore, postharvest fish handling, processing, preservation, packaging, storage, and transportation require particular care to maintain the quality and nutritional attributes of fish, thus avoiding waste and losses. Preservation and processing techniques can reduce the rate of spoilage, thus allowing fish to be distributed and marketed worldwide. Preservation and processing techniques involve temperature reduction (by chilling and freezing), heat treatment (by canning, boiling, and smoking), reduction of water content (by drying, salting, and smoking), and changing the storage environment (by packaging and refrigerating). However, fish can also be preserved, sold, and distributed using a wider range of other methods, such as in live form, and as various products destined for food or non-food uses. Technological development in food processing and packaging is ongoing in many countries, with increases in efficient, effective, and lucrative utilization of raw materials and innovation in product diversification. Moreover, the expansion in the consumption and commercialization of aquatic products in recent decades has been accompanied by growing interest in food quality, safety, nutrition, and waste reduction. As for food safety and consumer protection, increasingly stringent hygiene measures have been adopted at national and international trade levels.

The share of global aquatic products utilized for direct human consumption has significantly increased in recent decades, from 67% in the 1960s to 87%, that is, more than 146 million tons in 2014. The remaining 21 million tons of products were destined for non-food products, of which 76% (15.8 million tons) was processed into fishmeal and fish oil in 2014, while the rest being largely utilized as fish for ornamental purposes, aquaculture (fingerlings, fry, among others), bait, medicine, and raw materials for direct feeding in aquaculture, husbandry, and fur animal feed.

In 2014, 46% (67 million tons) of the food products for human consumption were live, fresh, or chilled fish, which were often the most preferred and high-priced forms of fish at most markets. The rest of the products for food purposes were in other different processed forms, with approximately 12% (17 million tons) dried, salted, smoked, or in other processed forms, 13% (19 million tons) in prepared and preserved forms, and 30% (approximately 44 million tons) in frozen form. Therefore, freezing was the main method of processing fish for human consumption, and it accounted for 55% of total processed fish for human consumption and 26% of total fish production.

The utilization of fish and, more significantly, the processing methods vary by continent, region, country, and even within a country. Latin American countries produce the highest percentage of fishmeal. In Europe and North America, more than two-thirds of fish for human consumption is frozen and prepared and preserved. Africa's proportion of cured fish is higher than the world average. In Asia, most fish products are still sold as live or fresh. Live fish is particularly appreciated in Southeast Asia and the Far East (especially in China) and in small markets in other countries (mainly among immigrant communities in Asian countries). Processing live fish for trade and consumption has been practiced in China and other countries for more than 3000 years. Commercialization of live fish has been enhanced in recent years as a result of technological development, improved logistics, and increased demand. However, live fish marketing and transportation are challenging due to strict sanitary regulations and quality requirements. In some areas in Southeast Asia, commercialization and trade of live fish are not formally regulated; instead, they are conducted in a traditional way. However, in the EU, market commercialization and trade of live fish have to comply with requirements, and animal welfare should be considered during transportation as well.

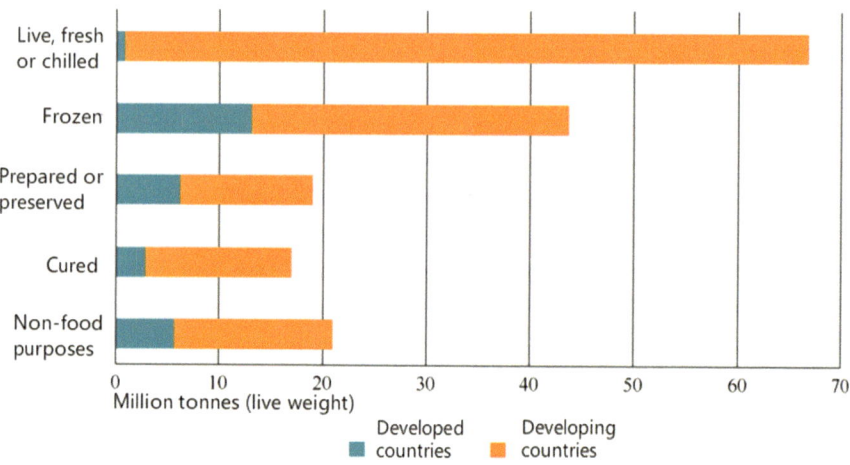

Fig. 2.16 Global fish utilization in 2014 (FAO 2016)

In recent decades, major innovations in refrigeration, ice making, and transportation have allowed a flexible distribution of fish in fresh and other forms. As a result, in developing countries the share of frozen fish for human consumption increased from 3% in the 1960s to 11% in the 1980s and 25% in 2014 (Fig. 2.16). Meanwhile, the share of prepared or preserved fish also grew (from 4% in the 1960s to 9% in the 1980s and 10% in 2014). However, despite technological development and innovation, many countries, especially those with less-developed economies, still lack adequate infrastructure and services, such as reliable electricity supply, potable water, roads, ice plants, cold rooms, refrigerated transport, and appropriate processing and storage facilities. Throughout the world, postharvest fish losses, which mostly occur in fish distribution chains, are a major concern, with an estimated 27% of landed fish being lost or wasted between landing and consumption. Globally, if discards prior to landing are included, fish losses and waste amount to 35% of landings, with at least 8% of fish discarded back into the sea.

Congested market infrastructure can also limit the marketing of fish products. The abovementioned deficiencies, together with established consumption habits, mean that fish products in developing countries, once harvested or landed, are commercialized mainly as live or fresh (representing 53% of fish destined for human consumption in 2014), or preserved using traditional methods, such as salting, drying, and smoking. These methods remain prevalent in many countries, especially in Africa and Asia. In developing countries, cured forms (dried, smoked, or fermented) represent 11% of all fish for human consumption. In many developing countries, processing techniques involve less sophisticated methods of transformation, such as filleting, salting, canning, drying, and fermenting. These labor-intensive methods support the livelihoods of many people in coastal areas, and they will probably remain important components of rural economies. However, in the last decade, fish processing techniques have also evolved in many developing countries, no longer limited to gutting, heading, or slicing and, instead, expanding to more advanced value-added techniques, such as breading, cooking, and individual quick-freezing, in consideration of the commodity and market value. Processing technique developments are driven by demand in the domestic retail industry, shifts in farmed species, outsourcing of processing, and more business contacts between producers in developing countries and firms abroad.

In recent decades, the aquatic food sector has become more heterogeneous and dynamic. Supermarket chains and large retailers are increasingly the key players in setting product requirements and influencing the expansion of international distribution channels. Processing is more intensive, geographically concentrated, vertically integrated, and linked to global supply chains. Processors are becoming more integrated with producers to enhance the combination of various product forms, obtain better yields, and actively respond to evolving quality and safety requirements in importing countries. The outsourcing of processing activities at the regional and global levels is playing a more significant role, with more countries involved, although the extent depends on species, product form, and labor and transportation costs. For example, whole frozen fish from European and North American markets are sent to Asia (especially to China, as well as other countries, such as India, Indonesia, and Vietnam) for filleting and packaging and then reimported. Further outsourcing of production to developing countries might be constrained by demanding hygiene requirements and growing labor costs in some countries, particularly in Asia, and transportation costs. All these factors might lead to changes in distribution and processing practices, finally resulting in increased fish product prices.

In developed countries, most fish for human consumption is a frozen product or in a prepared or preserved form. The proportion of frozen fish rose from 25% in the 1960s to 42% in the 1980s and reached the highest record of 57% in 2014. The proportion of prepared and preserved fish has remained stable, at 27% in 2014. In developed countries, innovations in value addition, together with changes in dietary habits, call for fish products mainly as fresh, frozen, breaded, smoked, or canned and as ready and/or portion-controlled, uniform-quality food. In addition, 13% of the fish products in developed countries destined for human consumption were in dried, salted, smoked, or other cured forms in 2014.

A significant, but declining, proportion of global fisheries production is processed into fishmeal and fish oil, thereby contributing indirectly to human consumption when these fish are used as feed in aquaculture and livestock feeding. Fishmeal is the crude powder obtained after milling and drying fish or fish parts, while fish oil is usually the clear brown/yellow liquid obtained through pressing cooked fish. These products can be produced from whole fish, fish remains, or other fish by-products resulting from processing. Many different fish species are used for fishmeal and fish oil production, and most of them are oily fish, such as anchovy, the primary species used for fishmeal. El Niño affects the anchovy harvest, and stricter management measures have reduced catches of anchovy and other species that are usually used as the raw materials for fishmeal. Therefore, fishmeal and fish oil productions fluctuate according to changes in the production of these species. Fishmeal production peaked in 1994 at 30.1 million tons (live weight equivalent) and has followed an oscillating and overall declining trend since then. In 2014, fishmeal production was 15.8 million tons due to reduced anchovy catches. Owing to the growing demand for fishmeal and fish oil, in particular, from the aquaculture industry, coupled with high prices, a growing proportion of fishmeal is being produced from fish by-products, which used to be often discarded. Unofficial statistics note that the contribution of by-products to the total production of fishmeal and fish oil produced is approximately 25% to 35%. With no additional expected raw material from whole fish (in particular, pelagic fish), any increase in fishmeal production will require recycling by-products, which may generate possible impacts on fishmeal composition.

The more advanced the processing level of fish products within a supply chain, the more the quantities of offal and other by-products, which may constitute up to 70% of the weight of fish and shellfish after industrialized processing. Fish by-products are not usually available at the market owing to low consumer acceptance and limited use due to demanding sanitary regulations. Fish by-products, including waste, used to be considered low value and used as feed for farmed animals or discarded. Over the past two decades, the utilization of fish by-products has gained attention because they can represent a significant additional source of nutrition. In some countries, the utilization of by-products has developed into an important industry. Heads, frames, and fillet cut-offs can be used directly as food or turned into products for human consumption, such as fish sausages, fish cakes, gelatin, and sauces. Other fish by-products are used as direct feed for aquaculture animals and livestock, pet food or feed for fur animals, fish silage, and fertilizers. Shark cartilage is utilized in many pharmaceutical preparations through reduction to powder, creams, and capsules, as are other shark parts, e.g., ovary, brain, skin, and stomach. Fish collagens are utilized for making cosmetics, and gelatin extracted from collagen is utilized in the food processing industry.

The internal organs of fish are an excellent source of specialized enzymes. Various proteolytic fish enzymes are extracted, e.g., pepsin, trypsin, chymotrypsin, collagenases, and lipase enzymes. As a good source of collagen and gelatin, fish bones are an excellent source of calcium and other minerals, such as phosphorus, that can be added to food, feed, or as supplements. Fish skin, particularly from larger fish, provides gelatin as well as leather for use in clothing, shoes, handbags, wallets, belts, and other items. In addition, shark teeth can be made into handicrafts.

The shells of crustaceans and bivalves are important by-products. Their efficient utilization is important due to the great quantity from increased production and increased processing and to the slow natural degradation rate of shells. Chitosan extracted from shrimp and crab shell has shown a wide range of applications, such as water treatment, cosmetics and toiletries, food and beverages, agrochemicals, and pharmaceuticals. Crustacean waste produces pigments (carotenoids and astaxanthin) for use in the pharmaceutical industry. Collagen can be extracted from fish skin, fins, and other processing by-products. Mussel shells can provide calcium carbonate for industrial use. In some countries, oyster shells are a raw material used in building construction and production of quicklime. Shells can also be processed into pearl powder and shell powder. Pearl powder is used in medicine and cosmetic manufacturing, and shell powder (a rich source of calcium) serves as a dietary supplement in feeding livestock and poultry. Fish scales are processed as fish silver, a raw material in medicines, biochemical drugs, and paint manufacturing.

In addition to the abovementioned products, in 2014, approximately 28.5 million tons of seaweed and other algae were harvested for direct consumption or further processing for food (traditionally in Japan, Korea and China) or for fertilizer, pharmaceuticals, cosmetics, and other purposes. Many seaweed-flavored foods (including ice cream) and drinks are being launched, with the Asia and Pacific region as the main market, with increasing interest from Europe and America. However, seaweed is characterized by a highly variable composition, depending on species, collection time, and habitat. More research is being conducted to explore the use of seaweed as an alternative to salt. Procedures are being developed for the industrial preparation of biofuel from fish waste and seaweed.

2.6.2 Status of the Utilization of Fishery By-Products

Globally, approximately 70 million tons of fish are processed by slicing, freezing, canning, and curing, which gives rise to by-products and waste in processing. For example, in the fish fillet industry, the final products usually only account for 30%–50%. The global production of all kinds of tuna was 4.76 million tons (live) in 2011, while the weight of canned tuna was approximately 2 million tons. In the tuna canning industry, up to 65% of raw materials turn out to be solid waste or by-products, which is usually composed of heads, bones, and so on. Additionally, 50% of raw materials in the tuna fillet industry are solid waste or by-products. Global salmon aquaculture production in 2011 was approximately 1.93 million tons, most of which are filleted, while some of which are sold after being smoked. The production rate of salmon fillets is 55%. A large proportion of farmed tilapia (global production at approximately 3.95 million tons in 2011) is processed as fillets, with a production rate between 30% and 37%.

2.6.2.1 Utilization of By-products for Food

Iceland and Norway have a long history in processing cod by-products. In 2011, Iceland exported 11,540 tons of dried cod heads, and the primary exporting area was Africa. Norway exported 3, 100 tons of processed by-products. Cod eggs can be eaten fresh after heat treatment, processed and canned, or processed into fish seed gum for sandwich sauce. Additionally, its liver can be canned or processed into cod liver oil.

The tuna industry has made significant progress in using by-products to produce products for human consumption. Thailand is the largest canned tuna producer in the world, with an annual export of approximately 500,000 tons. The raw materials are from domestic landings and 800,000 tons imported fresh or frozen tuna, of which

32–40% are processed into canned tuna. The dark parts of the tuna (10–30%) are canned or packaged into pet food. A by-product company in Thailand produces approximately 2000 tons of tuna oil a year, which can be further refined for human consumption. The fully refined tuna oil contains 25–30% DHA and EPA, an important ingredient in foods such as yoghurt, milk, baby formula, and bread.

Snacks made of tilapia skin are very popular in Thailand and Philippines. After frying in oil, the scale-removed tilapia skin can be processed into snack food. In some countries, scraps and heads, by-products from the fish fillet industry, are used to make soups and ceviche. There is equipment for removing fish bones, and boneless meat can be used as the base material for fish cakes, fish sausages, fish balls, and fish sauce.

2.6.2.2 Utilization of By-products for Feed

Because of the increasing global demand for fishmeal and fish oil, their prices have increased, and they are no longer low-value products. Pelagic fish are increasingly used directly as food, rather than as fishmeal. The implementation of rigid management measures, such as strict restrictions on the fisheries for raw materials for feed and improved rules and controls, has increased the price of fishmeal and fish oil. Hence, the proportion of fishmeal made from by-products increased from 25% in 2009 to 36% in 2010.

Thailand, Japan, and Chile are the three major producers of fishmeal from by-products. The International Fishmeal and Fish Oil Association estimates that the aquaculture industry used 73% of the fishmeal produced in 2010, which indirectly contributed to food production. As for fish oil, approximately 71% is used as aquaculture feed and 26% for human consumption.

2.6.2.3 Utilization of By-products for Health Supplements and Bioactive Ingredients

Long-chain polyunsaturated fatty acids, EPA, and DHA are probably the most commercially successful marine lipids derived from fish oil. Despite a late start in 2000, the market for omega-3 has grown rapidly. According to a market survey, the global demand for the ingredients in omega-3 was valued at $1.595 billion in 2010. The pharmaceutical and food industries use gelatin as an ingredient to improve product performance in texture, elasticity, consistency, and stability. In 2011, global gelatin production was approximately 348,900 tons, of which 98%–99% came from the skin and bones of pigs and cattle and approximately 1.5% came from fish and other sources. The market price of fish gelatin tends to be 4–5 times higher than that of mammalian, and it is only used for producing halal and kosher foods. Due to its rheological properties (in terms of physical consistency and flow), gelatin from warm water fish can be a substitute for the outer coatings of some food and medicine. Gelatin from cold water fish can be used to solidify and freeze food.

Chitin and deacetylated chitosan are widely used in food processing and pharmaceutical, cosmetic, and industrial production. Chitosan can be found in shrimp shells. A former industry forecast showed that the global market for chitin and chitosan in 2018 would be 118,000 tons. The unit structure of glucosamine and chitosan can be used for nutrition and medicine. Glucosamine and chondroitin sulfate can be used together to improve articular cartilage health as well as the ingredients in food and beverages. Among the aquaculture countries, China, Thailand, and Ecuador have established chitin and chitosan industries.

2.7 Modern Recreational Fisheries

As an emerging fisheries industry, recreational fisheries rapidly developed in the late twentieth century. Estimates suggest that approximately 10% of the population in developed countries are employed in recreational fisheries, and the number of people employed in recreational fisheries worldwide may exceed 140 million. A study outlining the value of ecosystem-based marine recreation estimates that the number of people employed in marine recreational fishing is 58 million. Millions of jobs depend on recreational fisheries and related expenses, which generate billions of USD in revenue each year. The recreational fishing industry is the most developed in the USA and European countries, and it is estimated that there are at least 60 million and 25 million casual anglers in the USA and Europe, respectively. In Europe, 8–10 million people are expected to be employed in recreational fishing in marine waters. Similarly, approximately 10% of Central Asia's population is expected to work in recreational fisheries.

During the "Twelfth Five-Year Plan" period, China's recreational fisheries showed an accelerated development rate of their rich content, industrial integration, and field expansion. In 2015, 380,000 entities were involved in recreational fishery operation, and they received more than 120 million tourists, with an output value exceeding 50 billion RMB. During the "Thirteenth Five-Year Plan" period, China has attached greater importance to developing the recreational tourism industry, and market demand is growing. Some experts predict that in the next 20 years recreational tourism in China will receive over 8 billion tourists, an explosive amount of growth, which will provide a huge potential for the development of recreational fisheries. As one of the five major industries of modern fisheries in China, recreational fisheries have been officially included in medium- and long-term fisheries development in China. Moreover, recreational fisheries provide a good niche to promote supply-side structural fisheries industrial reform, an important opportunity for fishermen to become employed and paid with additional income, and a significant means to alleviate industrial poverty as well.

2.7.1 Concept and Form of Recreational Fisheries

2.7.1.1 Concept of Recreational Fisheries

In the 1960s, recreational fisheries began to flourish in economically developed coastal countries and regions, such as the USA and Japan, and recreational fisheries have developed and become mature with social progress and economic development. Different scholars hold different views on the concept or definition of recreational fisheries. In the USA, recreational fisheries are considered recreational activities based on fishery resources, fitness-oriented activities, land- and water-based sport fishing, recreational gatherings, and home entertainment. These activities are often called recreational fisheries or sport fisheries to distinguish them from commercial fishing activities. Activities such as fishing village custom tourism, fishing village cultural leisure, and ornamental fisheries are not included in recreational fisheries. The definition of recreational fisheries in the USA and Western countries is very narrow.

However, Chinese fishery economists have various explanations and definitions for recreational fisheries based on the various purposes of fisheries activities. First, they maintain that the recreational fisheries industry is an industry operated in accordance with market rules, which organically combines natural resources, fishery resources, modern or traditional fishing gear and methods, fishing facilities and sites, fishermen work and home sites, and cultural resources of fishing villages with recreational fisheries activities, such as tourism, fishing entertainment, science education, fisheries expositions, and so on. Second, some economists argue that recreational fisheries belong to fisheries activities with extensive public participation, which is aimed at both physical and mental relaxation. Finally, some believe that recreational fisheries belong to industrial activities, which combine modern fisheries with leisure, tourism, sightseeing, and marine knowledge education. In recreational fisheries, the optimal allocation and rational use of fishery resources, environmental resources, and human resources can be obtained to achieve the combination and transfer of first, secondary, and tertiary industries for creating greater economic and social benefits.

2.7.1.2 Styles of Recreational Fisheries

Recreational fisheries are characterized by an organic combination of fisheries activities such as fishing, aquaculture, shell and algae collection, aquatic products trade, fish viewing, fish knowledge and fish product tasting with transportation, tourism, cooking, entertainment, and science education. Therefore, recreational fisheries have various modes.

Modern recreational fisheries can be divided into the following modes. First are the sport activities based on fishing. The second are the fishing activities, leisure experiences, and sightseeing that visitors can directly participate in, such as collecting shells, as well as the unique recreational fisheries tour, which was developed utilizing the unique characteristics of fishery resources and abundant resources. The third is culture of fish eating. The fourth is fisheries education and cultural appreciation, which is conducted in aquariums, fisheries fairs, and various exhibitions, with the purpose of education and technology popularization.

Recreational Angling

Recreational angling refers to recreational fishing activities that focus on fishing while integrating fishing with entertainment, fitness, and cooking. Recreational angling can be conducted in professional fishing parks and well-equipped fishing farms, where professional marine aquaculture cages and seawater and freshwater aquaculture ponds are available, and they are equipped with facilities for farming various seawater and freshwater fish for recreational fisheries. Recreational fishing can be divided into sea fishing, pond fishing, and cage fishing.

Recreational Fisheries with Experiences and Sightseeing

With the marine natural resources of fishing ports, shallow seas, and island reefs, a sea tourism base is established, where tourists enjoy sea fishing, intertidal collection, seascape sightseeing, and sea sports. Fishermen-house vacations provide tourists with a typical recreational fisheries experience. With fishing facilities, such as fishing boats, fishing gear, fishing village facilities, and fishing skills, a fishing village provides tourists with the opportunity to directly participate in traditional fishing, such as net setting and collecting, shrimp capture, cage capture, and sea fishing. Tourists can board fishing boats, set fishing nets, eat seafood, live in fishermen's houses to experience the life of a fishermen, and enjoy fishing and the customs of a fishing village. In addition, tourists can also collect shells at the beach, catch fish in a pond, participate in campfire parties, weave fishing nets, and cast feed, activities that tourists would enjoy participating in.

Some reservoirs and lakes are rich in unique fish products, such as silverfish in Taihu Lake and hairy crabs in Yangcheng Lake. Unique fish products tend to be the highlights of recreational fisheries. Some reservoirs and lakes are not only rich in unique fishery resources but are also unique in the operation of fisheries activities, which attract considerable tourists. For example, Qiandao Lake in Zhejiang Province is a wide-open lake, with beautiful scenery and abundant plankton. This lake is suitable for artificial release of such fish as bighead carp and white carp, which has facilitated the appreciation of a distinctive carp head eating culture. Qiandao Lake is a deep water body with diverse aquatic plant and animal species, giving rise to difficulties in fishing production. To facilitate fishing production at Qiandao Lake, combined fishing technologies defined as "barrier, driving, gill net, and drap net" have been introduced its "fishing with giant nets" as the recreational fishing activities have been introduced, which is now well-known both domestically and abroad.

Ornamental Fisheries

The ornamental fisheries industry integrates science education with entertainment, educating tourists with knowledge of various fish species in fishing entertainment museums, fishermen museums, marine museums, fisheries museums, fishing boat museums, aquariums, and fishing village museums. Ornamental fisheries are closely linked to the development of the aquarium industry and ornamental fish industry. By improving and optimizing the industrial structure and establishing ornamental aquariums in public places, we can improve the economic efficiency of the fishery industry and meet market demands as well. In addition, the development of ornamental fisheries also contributes to cultivating human moral values by loving nature and cherishing life.

Fish Eating Culture

Areas with convenient transportation can facilitate the development of a market for aquatic products trading. A large-scale market with excellent service and management can attract wholesalers and suppliers of aquatic products. A market with a variety of aquatic products tends to attract as many visitors as a museum does. Aquatic products are inseparable from dietary culture. "Fresh, live and tasty" fish has become characteristic of the fish eating culture, facilitating a dietary culture appreciation industry providing tourists with opportunities to taste, enjoy, and purchase in coastal areas. This industry can provide tourists with opportunities to participate in a variety of activities, such as enjoying sea breezes, watching ocean waves in a harbor, listening to fishermen's songs, eating seafood, enjoying evening views of a fishing port, visiting aquatic products exhibitions, attending trade fairs, participating in fishery industry development forums, and enjoying seafood festivals and marine cultural forums.

Educational Recreational Fisheries

In fishing ports and fishing villages with long histories, fishermen, for generations, have taken up activities such as weaving nets at the seaside, sailing, and fishing, which has precipitated a rich cultural heritage. Fish cultural displays can be integrated with museum displays and marine culture, by dividing exhibition areas according to the themes of time, species, and history. A time-themed exhibition hall can feature fishing gear from different periods and furniture in the style of fishing villages. A fish species-themed exhibition hall can display various fish specimens and introduce the biological and ecological characteristics, food values, and cultural legends of fish species to enhance tourists' understanding of the fisheries industry. Historical event-themed documents, pictures, and videos can be displayed in a cinema style. Sightseeing areas where fisheries culture is displayed should take the form of fishing village architecture and traditional style, fully discover the cultural connotations of a fishing village, and properly integrate modern fisheries technologies with folk customs.

Recreational fisheries are a typical hybrid industry that can be developed in many ways. Fishing, entertainment, vacationing at a fishermen's house, fishermen life experiences, sea activity experiences, eating and shopping, tasting seafood, fisheries trade, fisheries culture and recreation, island eco-tourism, sport fishing, and yacht sports can all be introduced in the recreational fisheries industry.

2.7.2 Development of Recreational Fisheries

Recreational fisheries originated in the USA. At the beginning of the nineteenth century, some fishing clubs were established along the Atlantic coast of the USA, and fishing at the clubs was different from commercial fishing. A fishing club provided individual tourists or families with recreational fishing activities to relax along lakes, rivers, or inshore waters. By the early twentieth century, recreational fisheries were solely recreational activities for fishing enthusiasts. In the 1950s, living standards were greatly enhanced because of rapid economic development, fewer working hours, and more free time. Tourism and recreational activities were increasingly favored; therefore, fishing and recreational activities in the USA rapidly developed at that time. In the 1960s, recreational fishing activities emerged in the Caribbean Sea and expanded to Europe and the Asia Pacific.

Japan. In the 1970s, Japan proposed a development strategy of "multi-use of the ocean." Japan took various measures to implement this strategy, such as constructing artificial reefs on the coast, establishing artificial fishing grounds, improving the environment of fishing ports and fishing villages, and vigorously developing recreational fisheries. With increased income and more free time, more Japanese have preferred to visit coastal areas around fishing ports since 1975. As a form of healthy entertainment, sport fishing developed quickly. In 1993, the number of tourists involved in sport fishing in Japan reached 37.29 million, accounting for 30% of the country's total population, while the number of people employed in the sport fishing guided tour industry reached 24,000. The development of the sport fishing industry greatly promoted the economy in Japanese fishing villages, optimized the structure of the Japanese fishery industry, and promoted sustainable fisheries development. In 1990, Chinese Taipei implemented a vessel reduction policy to adjust the fisheries industrial structure and encouraged the development of recreational fisheries along coastal areas and ports, which promoted the development of local recreational fisheries.

Chinese Taipei. Due to the decline of inshore living marine resources, the limited development of deep-sea fishing, and the shortage in the labor force, fisheries development in Chinese Taipei has faced various difficulties in recent years. Since 1998, the Fisheries Bureau of Chinese Taipei has increased investment in recreational facilities at six fishing ports, including Keelung, and developed marine and terrestrial recreation centers to promote diversified employment of fishermen. The facilities at the recreational fishing center include marina fishing and sightseeing, a fisherman's wharf, seafood court, sea fishing club, sea sightseeing park, children's garden, hotel, and tourism services. In the second half of 1998, 99 seaports were open to recreational fisheries, and more than 700 fishing boats were approved for recreational fisheries. To promote the development of recreational fisheries, attract more tourists and urban residents to fishing ports and fishing areas for sightseeing, and promote the economy in fishing areas, outlet centers for aquatic products were established at the main fishing ports. Thus, tourists can eat and purchase delicious fish products while enjoying a view of the fishing ports and fishing villages. The fishing ports and fishing areas in Chinese Taipei, which combine production, sales, entertainment, and sightseeing, have revitalized the coastal and inshore fisheries in Chinese Taipei.

Mainland China. Mainland China is located in the northern temperate zone and subtropical zone, which facilitates a longer season for recreational fisheries. The southeastern coast is especially suitable for fishing for 8–9 months every year. The superior environmental and biological resources have laid a good foundation for the development of recreational fisheries. In the mid-1990s, recreational fisheries began to rapidly develop in large cities and coastal cities in China. Huairou and Fangshan districts in the suburbs of Beijing established a recreational fisheries visiting attraction, integrating sightseeing, fishing, and eating, along with the development of rainbow trout aquaculture, which turned out to be a great success by achieving high profits. Sanhe County in Langfang City of Hebei Province produced more than 1100 tons of commercial fish annually, of which more than one-third were used for sport fishing. The income from sport fishing accounted for approximately 50% of the total fisheries revenue of the county. Changhai County in Dalian City of Liaoning Province, with a geographical advantage, held a fishing festival, which attracted many domestic and foreign tourists to participate in fishing competitions and promoted the local economy. In the western region, Quxian County of Sichuan Province, taking advantage of the landscape scenery on both sides of the Qujiang River, developed new tourism activities. In Cangzhou City of Fujian Province, a coastal area located on the southeastern coast of China with beautiful scenery, surrounding mountains, rivers, a 680-km coastline, and numerous islands, recreational fisheries were also developed quickly.

USA. The US fishery industry can be divided into two components, that is, commercial fisheries and recreational fisheries. East of the Atlantic Ocean and west of the Pacific Ocean, with a total coastline of 22,680 km, the USA has superior water resources. The country has a densely interconnected inland water system, with numerous lakes and reservoirs. However, due to a high production cost, the economic benefits of US commercial fisheries are not high. However, recreational fisheries are not only economically efficient but also highly developed, and they have developed into a pillar industry in modern fisheries. In the early twenty-first century, recreational fisheries received approximately 35.2 million tourists/year in the USA, who spent $37.8 billion on recreational fishing. If recreational fisheries are treated as an enterprise, they rank 13th on the Fortune 500 with their significant revenue. Marine recreational fisheries are also highly developed. In 2001, the number of recreational anglers increased to 17 million, and they made more than 86 million saltwater fishing trips. In the USA, the main species for recreational fishing are bluefish (*Pomatomus saltatrix*), red drum (*Sciaenops ocellatus*), and Spanish mackerel (*Scomberomorus commerson*), among others.

To protect marine fishery resources, the US federal government and the state government stipulate that those who fish in public waters should apply to the government fisheries management department and purchase a fishing license annually. The license fee varies from state to state. The fees paid by anglers for the fishing license are mainly used for the construction of fishing areas and resource conservation. Both federal and state governments stipulate that anglers with fishing licenses are not allowed to fish whenever and wherever they like. According to the law, in addition to being restricted to one person and one hook, anglers are also restricted by regulations on the species, amount, and size of the catch. According to Texas state law, 24 fish species are allowed to be fished, including sea bass and tarpon. In Wisconsin state, one fisherman is allowed to take away two fish at a time. However, in New York state, catches should not exceed 25 kg in weight and should be less than five fish in quantity.

Former US President Bill Clinton signed Executive Order 12962 on June 7th, 1995, stating the importance of recreational fisheries to society, culture, and economy and requiring that the federal government improves the quantity, function, and sustainable production of US aquatic resources to increase employment opportunities in recreational fisheries.

The fundamental policies adopted by the US government to protect and support the sustainable development of recreational fisheries are mainly listed as follows. (1) The first is to protect, enhance, and restore the resources of important fish species for recreational fisheries and their habitats. (2) The second policy encourages the use of engineering facilities, such as artificial reefs. (3) The third policy provides public education. The government should support, develop, and implement education to raise public awareness of conserving marine resources and improving recreational fisheries. (4) The fourth policy establishes and encourages partnerships between the government and civil society organizations. Through bilateral cooperation, the management and resource protection of recreational fisheries should be enhanced, and opportunities for the development of recreational fisheries should also be enhanced.

The US government has invested heavily in the development of recreational fisheries. The government attaches great importance to the conservation of fishery resources and their habitats. The federal government has set up a large number of management and research institutions throughout the country to conduct extensive and in-depth research on fishery resources biology, ecology, and sport fishing activities.

2.7.3 Management and Development of Recreational Fisheries

2.7.3.1 Challenges of Recreational Fisheries

The recreational fisheries industry is a developed industry in most developed countries, while it is developing quickly in developing countries. This industry involves a large number of individuals. Considering the number of employees, production, and social and economic correlation, recreational fishing is a major industry, knowledge which is universally accepted. However, in many recreational fisheries, this awareness has not led to improvements in management. The influence of recreational fisheries development on the livelihoods of full-time fishermen, environmental protection, and aquatic biodiversity is receiving increasing attention.

However, sometimes recreational fisheries tourists have a negative impact on specialized small-scale fishermen and artisanal fishermen in open fishing areas and public fishing grounds. There are also debates and issues related to the dangers of recreational fisheries, such as the use of harpoons to catch grouper in the Mediterranean Sea, the Australian coastal areas, and the eastern Red Sea. In addition, the capture of some species in recreational fisheries, such as eye-dragon lobsters, together with the impact of commercial fisheries and other stresses (such as pollution), can lead to the significant decline of some fish species.

Currently, recreational fisheries tourists can go farther into open seas, fish with techniques, and be equipped with some fishing equipment that enables them to achieve a fishing capacity similar to that of commercial fishermen. Recreational fishery species used to be exploited only by commercial fisheries, which has led to conflicts in some areas between recreational and commercial fisheries. Targeted fishing and using the same type of gear and equipment, such as berthing, also bring about competition between recreational fisheries and small-scale commercial fisheries along the coast. Other special recreational fisheries, such as those targeting salmon, marlin, sailfish, and swordfish, fish regularly in specific regions and seasons, and this production accounts for a large proportion of the total production. However, notably, recreational fisheries have actively promoted catch-and-release activity. Fish caught in fishing competitions in recreational fisheries are generally released unless the caught fish are recorded.

Many recreational fisheries are highly selective. Usually, large-sized fish are targeted by recreational fisheries. However, the capture of larger-sized fish, which have a longer lifespan, harms the potential reproduction of a fish species. Larger-sized female fish lay more eggs and have longer spawning periods (therefore being more adaptable to a changing environment). The fry have a higher survival rate. Capture of these larger-sized fish affects spawning. The age, size, and species are affected by fish stock density changes and interactions, which also influences the food chain and the structure and productivity of an ecosystem. All of these factors are assumed to be more relevant when these species are exploited simultaneously in both commercial and recreational fisheries.

2.7.3.2 Issues to be Considered for Recreational Fisheries

The sustainable development of recreational fisheries depends on the recognition of its multidisciplinary characteristics. Regardless of whether recreational fisheries stakeholders are allowed to promote successful conservation and management, there is an urgent need to integrate multiple disciplines, such as biology and social science, to provide information on the dynamics between the recreational fisheries industry and the ecosystem.

Therefore, interdisciplinary researchers should explore how to integrate sustainable recreational fisheries (including the biodiversity of aquatic animals in fishing areas) with commercial fisheries. Policy makers and managers for recreational fisheries development need to be informed about how to integrate sustainable recreational fisheries with commercial fisheries and potential negative effects (including coastal development, fish habitat degradation, pollution, and extreme weather) on such integration. In addition, recreational fisheries have important social impacts, and their fishing benefits need to match the investment in resource conservation.

The management of recreational fisheries needs to coordinate the conflict with wild fish conservation and ensure the sustainable development of marine animals and the marine ecosystem in which these animals live. Therefore, the management of recreational fisheries needs to follow the approaches adopted by most other fisheries industries, and these approaches are listed as follows: (1) identifying the resources which need to be managed, system conditions, and restrictions, (2) determining objectives, (3) evaluating management options, (4) selecting appropriate actions to achieve management objectives, (5) implementing actions and monitoring the results, and (6) evaluating this management and adjusting management based on the evaluation. There are many management methods for freshwater recreational fisheries, such as artificial release, bioremediation, prey proliferation, suppression of harmful species, selective fishing, innovation, and management of aquatic plants.

2.8 Global Fishery Products Trade

2.8.1 Role and Status of Global Fishery Products Trade in International Trade

Trade plays a major role in the fisheries and aquaculture sector as a job creator, food supplier, income generator, and contributor to economic growth and development and food and nutrition security. Fish and fishery products represent one of the most-traded segments of the global food sector, with approximately 78% of seafood products estimated to be exposed to international trade competition. For many countries and numerous coastal, riverine, insular, and inland regions, exports of fish and fishery products are essential to the economy. For example, in 2014, fish and

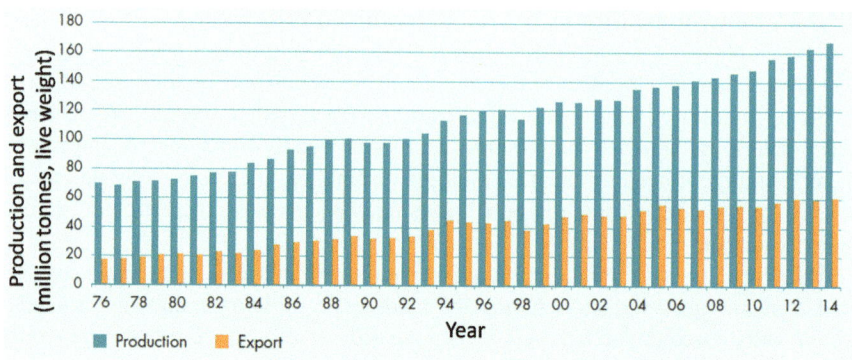

Fig. 2.17 Global fisheries production and export (FAO 2016)

fishery products accounted for more than 40% of the total value of traded commodities in Cabo Verde, the Faroe Islands, Greenland, Iceland, Maldives, Seychelles, and Vanuatu. Moreover, in 2014, global fisheries trade represented more than 9% of total agricultural exports (excluding forest products) and 1% of global merchandise trade in terms of the trade value.

World trade in fish and fishery products has significantly expanded in recent decades, increasing by more than 245% in terms of quantity from 1976 to 2014 and by 515% in terms of the trade of fish for only human consumption. The trade volume represented a significant share of total fisheries production, approximately 36% (live weight equivalent) exported in different product forms for human consumption purposes or non-edible purposes in 2014 (Fig. 2.17), reflecting this sector's degree of openness and integration into international trade. This share increased from 25% in 1976 to a peak of 40% in 2005. Since then, it has slowed, mainly because of reduced production and related fishmeal exports. If only trade of fish for human consumption is considered, its share in total fisheries production has increased continuously, reaching almost 29% in 2014. The global trade of fish and fish products has also significantly increased by value. The value of exports increased from $8 billion in 1976 to $148 billion in 2014, with an annual growth rate of 8.0% and a constant price growth rate of 4.6%.

2.8.2 Major Fishery Products Trading Countries

Table 2.18 indicates the major importers and exporters of fishery products. China is the largest fish producer and has also been the largest exporter of fish and fish products since 2002, although fish exports only comprise 1% of Chinese exports. China has become the third largest importer of fishery products due to the increasing quantities of imported fish since 2011. The growth of Chinese imports is the result of outsourcing the processing of other countries, which also reflects the growing

Table 2.18 Top 10 exporters and importers of fishery products (FAO 2016)

		2004 (million USD)	2014 (million USD)	Average annual growth rate percentage 2004–2014 (%)
Exporter	China	6637	20,980	12.2
	Norway	4132	10,803	10.1
	Vietnam	2444	8029	12.6
	Thailand	4060	6565	4.9
	USA	3851	6144	4.8
	Chile	2501	5854	8.9
	India	1409	5604	14.8
	Denmark	3566	4765	2.9
	Netherlands	2452	4555	6.4
	Canada	3487	4503	2.6
	Top 10 subtotal	34,539	77,802	8.5
	Other countries subtotal	37,330	70,346	6.5
World total		71,869	148,147	7.5
Importer	USA	11,964	20,317	5.4
	Japan	14,560	14,844	0.2
	China	3126	8501	10.5
	Spain	5222	7051	3.0
	France	4176	6670	4.8
	Germany	2805	6205	8.3
	Italy	3904	6166	4.7
	Sweden	1301	4783	13.9
	UK	2812	4638	5.1
	Korea	2250	4271	6.6
	Top 10 subtotal	52,120	83,446	4.8
	Other countries subtotal	23,583	57,169	9.3
World total		75,702	140,616	6.4

domestic demand in China for fish species that cannot be produced locally. However, in 2015, after years of growth, the Chinese fisheries trade slowed down, and exports decreased by 6% in USD (by 4% in RMB).

Norway, the second largest exporter, supplies diverse products, including products from farmed salmon, small pelagic species, and traditional whitefish. In 2015, Norway posted record export values, in particular, the export values for salmon and cod. In 2004, Vietnam overtook Thailand as the third largest exporter. Thailand has experienced a substantial decline in exports since 2013, mainly due to the reduced shrimp production caused by shrimp disease. Thailand's exports further declined in 2015 (by 14% in USD) mainly because of reduced shrimp production and lower prices of shrimp and tuna. Both Vietnam and Thailand have important processing

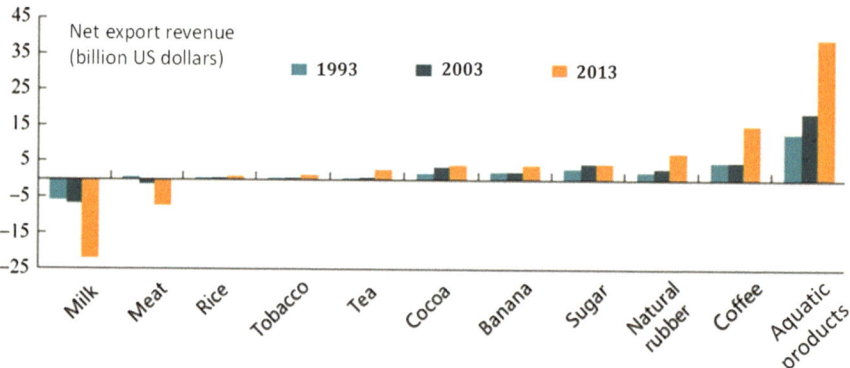

Fig. 2.18 Net exports of selected agricultural commodities by developing countries (FAO 2016)

industries, which significantly contribute to the national economy through job creation and trade.

The EU, the USA, and Japan are highly dependent on fishery products imports to satisfy domestic demand. In 2014, their combined imports represented 63% by value and 59% by volume. The EU is by far the largest market for imported fish, valued at $54 billion in 2014 ($28 billion if intra-EU trade is excluded). In recent years, Japanese fishery imports have declined, owing to a weaker currency, which has made imports more expensive. In 2015, its imports of fish and fishery products declined by 9% in USD to $13.5 billion, but increased by 4% in JPY. In 2015, US fishery imports reached $18.8 billion, decreasing by 7% compared to 2014.

One of the most important changes in trade patterns in recent years has been the growing share of developing countries in the fisheries trade and the corresponding decline in the share of developed economies. Developing economies, whose exports represented only 37% of world trade in 1976, saw their share rise to 54% of total fisheries export value in 2014. Meanwhile, their exports increased from 38% to 60% in volume (live weight) of total fisheries exports. Fisheries trade represents a significant source of foreign currency earnings for many developing countries, in addition to the sector's important role in enhancing income, employment, and food security and nutrition. However, its role varies considerably among developing countries and even within some specific regions. In 2014, exports of developing countries were valued at $80 billion, and their fisheries net export revenues (exports minus imports) reached $42 billion, higher than the combined revenues of other agricultural commodities (such as meat, tobacco, rice, and sugar) combined (Fig. 2.18). The fisheries industries in developing countries rely heavily on developed countries both as outlets for their exports and as suppliers of their imports for local consumption or for their processing industries.

Trade of fish and fishery products is largely driven by the demand from developed countries, who dominate most world fishery imports. In terms of volume (live weight equivalent), their share is very low at 57%, reflecting the higher unit value of the products they import. Their imports from capture and aquaculture originate from both developed and developing countries, giving many producers an incentive to produce, process, and export.

2.8.3 Global Fishery Products Consumption

The significant growth in capture and aquaculture production over the past 50 years, especially over the last 20 years, has enhanced the capacity of providing diversified and nutritious food. In terms of the daily global average, fish products provide only approximately 34 calories per capita. However, it can exceed 130 calories per capita in countries where there is a lack of alternative protein food and where a preference for fish has been developed and maintained (e.g., Iceland, Japan). The contribution of fish is more significant in terms of providing animal proteins, as 150 g of fish can provide approximately 50–60% of the daily protein requirements for an adult. Fish protein can represent a crucial component in the diets of some densely populated countries where total protein intake levels may be low. Many of these countries reveal heavy dependence on staple foods, with fish consumption becoming particularly important in improving the calorie/protein ratio. In addition, for people in those countries, fish often represents an affordable source of animal protein that may not only be cheaper than other animal protein sources, but preferred food and part of local and traditional recipes. For example, fish contributes, or exceeds, 50% of total animal protein intake in some small island states, as well as in Bangladesh, Cambodia, Ghana, Indonesia, Sierra Leone, and Sri Lanka. In 2013, at global level, fish accounted for approximately 17% of animal protein and 6.7% of all protein. Moreover, fish provided more than 3.1 billion people with almost 20% of their average per capita animal protein intake (Fig. 2.19).

Generally speaking, growth in the global supply of fish for human consumption has outpaced population growth over the past five decades, increasing at an average annual rate of 3.2% in the period 1961–2013, double that of population growth,

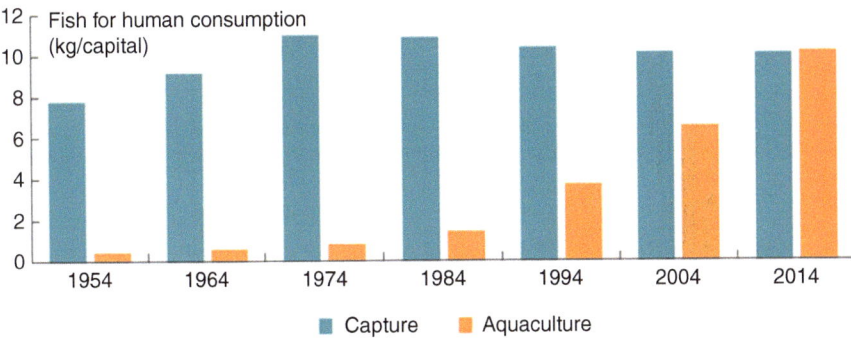

Fig. 2.19 Relative contribution of aquaculture and capture to the fish products for human consumption (FAO 2016)

Table 2.19 Total and per capita food fish supply by continent and economic grouping in 2013 (FAO 2016)

	Total food supply (million tons; live weight equivalent)	Per capita food supply/(kg/yr)
Global	140.8	19.7
Global (excluding China)	88.3	15.3
Africa	10.9	9.8
North America	7.6	21.4
Latin America and the Caribbean	5.8	9.4
Asia	99.0	23.0
Europe	16.5	22.2
Oceania	1.0	24.8
Industrialized countries	26.5	26.8
Other developed countries	5.6	13.9
Least-developed countries	11.1	12.4
Other developing countries	97.6	20.0
Low-income, food-deficit countries	18.6	7.6

resulting in increasing average per capita availability. World per capita apparent fish consumption increased from an average of 9.9 kg in the 1960s to 14.4 kg in the 1990s and 19.7 kg in 2013, and further growth is expected. In addition to the increased production, other factors that have contributed to rising consumption include reduced waste, better utilization, improved distribution channels, growing demand, population growth, rising incomes, and urbanization. International trade has also played an important role in providing more options for consumers.

The contribution of fish to nutritional intake varies considerably between and within countries and regions in terms of quantity and variety consumed per capita (Table 2.19). These dissimilarities in consumption depend on the availability and cost of fish and alternative food, as well as the accessibility of fishery resources in adjacent waters, disposable income, and socioeconomic and cultural factors, such as dietary traditions, eating habits, tastes, demand, seasons, prices, marketing, infrastructure, and communication facilities. Annual per capita apparent fish consumption can vary from less than 1 kg in one country to more than 100 kg in another. Variations may also be significant within a country, with consumption usually higher in coastal, riverine, and inland water areas.

Disparities in fish consumption also exist between more-developed and less-developed countries. Although annual per capita consumption of fishery products has grown steadily in developing regions (from 5.2 kg in 1961 to 18.8 kg in 2013) and in low-income, food-deficit countries (from 3.5 to 7.6 kg), it is still considerably lower than in more-developed regions. In 2013, per capita fish consumption in industrialized countries was 26.8 kg, while for all developed countries, it was estimated at 23.0 kg. A growing share of fish consumed in developed countries originates from imports, owing to steady demand and static or declining domestic fisheries production. In developing countries, fish consumption tends to be based on locally and seasonally available products, and the fish chain is driven by supply rather than demand. However, due to rising domestic income and wealth, consumers in emerging economies are experiencing a diversification of the fish products available owing to an increase in fishery imports.

Disparities among developed and developing countries exist due to the different contributions of fish to animal protein intake. Despite relatively lower levels of fish consumption, developing countries and low-income and food-deficit countries have a higher share of fish protein in their diets compared with developed countries and the world average. In 2013, fish accounted for approximately 20% of animal protein intake in developing countries and approximately 18% in low-income, food-deficit countries. This share has been increasing but has stagnated in recent years due to the growing consumption of other animal proteins. In developed countries, the share of fish in animal protein intake, after consistent growth since 1989, dropped from 13.9% in 1989 to 11.7% in 2013, while consumption of other animal proteins continued to increase.

2.8.4 Composition of World Trade of Fish and Fishery Products

Trade of fish and fishery products is becoming more complex, dynamic, and highly segmented, with greater diversification among species and product forms. An important share of fishery trade consists of high-value species, such as salmon, prawn, tuna, groundfish, bass, and bream. However, some high-volume, but relatively low-value, species are also traded in large quantities not only nationally but also at the regional and international level. For example, small pelagic fish are traded in large quantities, mainly consumed by low-income consumers in developing countries. However, emerging economies from developing countries have imported more high-value fish for their domestic consumption.

In recent decades, the dramatic expansion in aquaculture production has significantly contributed to the increased consumption and commercialization of species that used to be provided by capture, with farmed products representing a growing share of international fish trade. Despite recent improvements in trade classifications, international trade statistics do not distinguish between captured and farmed product origin. Hence, a distinction may be stated between captured and farmed products in international trade. Estimates indicate that aquaculture products represent between 20% and 25% of trade volume but 33% and 35% of trade value, indicating that an important industry segment is export-oriented and conducted by the producer of relatively high-value products destined for international markets. If only fishery products for direct human consumption are considered, this share increases to 26–28% of trade volume and 35–37% of trade value.

Owing to their high perishability, 92% of trade of fishery products in volume (live weight equivalent) consisted of processed products (i.e., excluding live and fresh whole fish) in 2014. Fishery products are increasingly traded as frozen food (40% of the total volume in 2014; 22% in 1984). Over the last four decades, prepared and preserved fish, including many value-added products, have doubled their share in total volume, increasing from 9% in 1984 to 18% in 2014. Although they are prone to spoilage, the trade in live, fresh, and chilled forms has increased due to consumer demand and innovation in chilling, packaging, and distribution technology, representing approximately 10% of global fishery products trade in 2014. Trade of live fish also includes ornamental fish, which are high in value but almost negligible in weight. In 2014, 78% of the exported volume consisted of products destined for human consumption. A large amount of fishmeal and fish oil is traded because, generally, the major producers (South America, among others) are distant from the main consumption markets (Europe and Asia).

The $148 billion in exports of fishery products in 2014 does not include an additional $1.8 billion represented by seaweed and other aquatic plants (62%), inedible fish by-products (27%), and sponges and corals (11%). Trade in aquatic plants increased from $0.1 billion in 1984 to more than $1 billion in 2014, with Indonesia, Chile, and South Korea as the largest exporters and China, Japan, and the USA as the leading importers. Owing to the increasing production of fishmeal and other products derived from fishery residues from processing, trade of inedible fish by-products has also surged, increasing from $90 million in 1984 to $0.2 billion in 2004 and $0.5 billion in 2014.

2.8.5 Trends in Global Trade of Fishery Products

2.8.5.1 Increasing Import and Export Trends in the Global Trade of Fishery Products

Due to the constraints of the labor force and rising production costs in developed countries, direct fishery production may be gradually reduced, but the demand for fishery products is still growing. The fisheries production technology in developing countries is constantly improving. To meet the needs of job enhancement and foreign exchange earnings, fisheries production may continue to grow, which will promote the development of global fishery products trade.

2.8.5.2 Technical Barriers

With improved living standards, people have greater demand for the safety and quality of fishery products. Some countries have established regulations on quality standards for fishery products, which have caused technical barriers in the trade of global fishery products. The standards for residues and feed additives for fishery products have significantly increased. Products that do not meet a standard are returned or destroyed on the spot. Labeling management has also been applied to some products, such as the "red label" in France and the "organic farmed fish label" in Ireland and Canada.

2.8.5.3 Traceability System

Implementing a traceability system means that all information, such as wholesalers, subpackaging, processing and breeding, and fishing vessels, can be traced to ensure the quality and responsibility of fishery products. Once a problem is detected in the quality of a fishery product, the production process can be traced and the cause of the problem can be defined. Traceability labeling can also be applied.

2.8.5.4 Ecolabel System

An ecolabel system is introduced to implement sustainable development, better utilization of fishery resources, and protection of the ecological environment. These systems aim to indicate that a fishery product with an ecolabel is not overfished nor a product that is harmful to other marine animals, such as marine mammals, turtles, and seabirds, thus promoting fisheries management. The FAO has developed a *Guide to Ecolabelling in the Marine Capture Industry*, which serves as a reference for internationally recognized, coordinated ecolabeling schemes and plays a guiding role in certification and designation.

2.9 Status and Trends of International Fisheries Management

2.9.1 Status of International Fisheries Management

The development of international fisheries management is, in fact, a process of continuously understanding the characteristics of fishery resources and improving fisheries management, as well as a process of continuously solving emerging problems in fisheries management. New concepts emerging in fisheries management are designed to address these new problems to ensure the sustainable and rational use of fishery resources. For example, the emergence of the concepts of "freedom of the high seas," "territorial waters," "exclusive economic zone," and "contiguous zone" reflects the needs for marine management and economic development at different times. In the late 1980s, the development and utilization of world fishery resources reached its peak, and some traditional fishery resources were declining. Under such circumstances, many new fisheries management concepts and measures were introduced to ensure the sustainable use of fishery resources. The emergence of new concepts itself also reflects the direction and trends of international fisheries management.

2.9.1.1 Responsible Fisheries

The concept of responsible fisheries was put forward at the Cancun Declaration, stating that "this concept encompasses the sustainable utilization of fishery resources in harmony with the environment; the use of capture and aquaculture practices which are not harmful to ecosystems, resources or their quality; the incorporation of added value to such products through transformation processes meeting the required sanitary standards; and the conduct of commercial practices to provide consumers access to good quality products." The essence of responsible fisheries is requiring people to be responsible for all activities related to fisheries to ensure the sustainable use of fishery resources. This concept is fully reflected in the two international documents *Conservation and Management of Straddling Fish Stocks and Highly Migratory Fish Stocks* and *International Conservation and Management Measures by Fishing Vessels on the High Seas.*

2.9.1.2 Precautionary Approach

The precautionary approach was proposed at the Technical Consultation Conference on High Seas Fishing held in September 1992. The precautionary approach is a kind of "result" produced under "uncertainty." The precautionary approach requires that the development of any new fishery or the expansion of an existing fishery should be determined after assessing the potential impacts on target and nontarget species. Adoption of the precautionary approach indicates that the management and utilization of fishery resources have shifted from maintaining its optimal use to sustainable resource use and from resource utilization to precautionary resource utilization.

2.9.1.3 Flag State Responsibility

In traditional international law, the flag state has exclusive jurisdiction over ships flying its flag, as reflected in the 1958 *Convention on the High Seas* and the 1982 *United Nations Convention on the Law of the Sea.* These conventions stipulate that there must be a real connection between the state and the ship and that a state exercises jurisdiction and control over the administrative, technical, and social matters of a ship flying its flag. When the international community discusses management of high seas fishery resources, the implementation of flag state responsibility has become one of the main means of fisheries resource conservation and management measures. Flag state responsibility has become the main content of three international fisheries conservation treaties, such as the *Agreement to Promote Compliance with International Conservation and Management Measures by Fishing Vessels on the High Seas.*

2.9.1.4 Design of Law Enforcement Mechanisms

The key to fisheries management is how to establish and design enforcement mechanisms. The *Agreement for the Implementation of the Provisions of the United Nations Convention on the Law of 10 December 1982 Relating to the Conservation and Management of Straddling Fish Stocks and Highly Migratory Fish Stocks* (referred to as UNFSA) emphasizes that regional or subregional fisheries management organizations should cooperate to ensure compliance and enforcement of conservation and management measures for straddling fish stocks and highly migratory fish stocks at regional or subregional levels. Regional fisheries management will serve as the main means of high seas fisheries resource management in the future. This assessment is reflected in the UNFSA and the Rome Declaration.

2.9.1.5 Regional Fishery Bodies

Regional fishery bodies play a critical role in the management of shared fishery resources. There are approximately 50 regional fishery bodies in the world. Only with the permission of member states can regional fishery bodies effectively play a role in fisheries management. Their role depends directly on the participation and political will of members.

With the development of *Agenda 21* to the *Code of Conduct for Responsible Fisheries*, the conservation and management of fishery resources has been combined with the protection of natural resources and the environment, and fisheries development has been combined with the world trading system and human health, safety, and welfare. At the same time, the formulation of national fisheries policy and regulations must take into account the comprehensive management of coastal areas and the needs and regulations of the WTO. The sustainable exploitation and utilization of fishery resources is the supreme goal of global fisheries management.

For a long time into the future, the development trends of world fisheries management are expected to further improve the measures and methods of fisheries management based on the *Code of Conduct for Responsible Fisheries*. For example, the *International Fisheries Action Plan*, adopted in February 1999, is a concrete action plan being implemented for responsible fisheries. Among them, the management of fishing capacity is attracting increasing attention. Excessive capacity actually leads to overfishing, the decline of marine fishery resources, reduced production potential, and decreased economic benefits. This action plan requires countries and regional organizations to establish strict, effective, fair, and transparent management of fishing capacity worldwide, which needs to be implemented step by step in stages through cooperation between governments or regional fisheries management organizations.

2.9.2 Nature and the Role of International Fisheries Organizations

An international fisheries organization is the general name for an agency jointly established by two or more countries or their civil organizations under specific agreement for the purpose of fisheries development, management, and cooperation. However, these organizations generally coordinate the relationship on fishery activities between states or national civil society, rather than the relationship on fisheries activities between certain fishery enterprises or between fishery operators.

In principle, international fisheries organizations can be divided into intergovernmental fisheries organizations and nongovernmental fisheries organizations. The former must involve national governments. However, the latter does not represent governments, and private fishing groups may participate.

International fisheries organizations can be divided into three categories by region: global, regional, and subregional. For example, the FAO Fisheries and Aquaculture Commission and the International Whaling Commission are global international fisheries organizations, while the Indian Ocean Fisheries Council is a regional international fisheries organization, and the Fishery Committee for the Eastern Central Atlantic (CECAF) is a subregional international fisheries organization. However, it is difficult to distinguish between regional and subregional international fisheries organizations, which are relative.

International fisheries organizations can also be divided into two categories characterized as either being affiliated with the FAO or non-affiliated with the FAO. The international fisheries organizations currently affiliated with the FAO include the Asia-Pacific Fisheries Commission (APFIC), Fishery Committee for the Eastern Central Atlantic (CECAF), Western Central Atlantic Fisheries Commission (WCAFC), General Fisheries Commission for the Mediterranean (GFCM), Indian Ocean Tuna Commission (IOTC), European Inland Fisheries and Aquaculture Advisory Commission (EIFAC), Commission for Small-Scale and Artisanal Fisheries and Aquaculture of Latin America and the Caribbean (COPPESAALC), and so on. The international organizations not currently affiliated with the FAO include Secretariat of the Pacific Community (SPC), South Pacific Regional Fisheries Management Organisation (SPRFMO), International Whaling Commission (IWC), Northwest Atlantic Fisheries Organization (NAFO), North-East Atlantic Fisheries Commission (NEAFC), International Commission for the Conservation of Atlantic Tunas (ICCAT), Inter-American Tropical Tuna Commission (IATTC), Commission for the Conservation of Southern Bluefin Tuna (CCSBT), International Pacific Halibut Commission (IPHC), and others.

According to the number of participating countries, international fisheries organizations can be divided into multilateral national fisheries organizations and bilateral national fisheries organizations. A case in point for multilateral national fisheries organizations is the Conservation Committee of Pollock resources in the Central Bering Sea, which consists of six countries: China, the USA, Russia, Japan, South Korea, and Poland. However, a case in point for bilateral national fisheries organizations is the China-Japan Joint Fisheries Commission.

2.9.3   Functions of International Fisheries Organizations

The mission of each international fishery organization is based on a signed agreement or convention. Generally, the functions can be summarized into three areas. (1) Investigation and research. An organization should engage in investigation and research of fishery resources and provide reports to members. (2) Consulting. An organization should provide advice to members as needed or upon request. (3) Management. An organization should carry out fisheries management based on agreed measures for the conservation and management of fishery resources, including joint enforcement.

Before the 1980s, the mission of international fisheries organizations focused on consultation. Generally speaking, within the waters under its jurisdiction, an organization was engaged in the following five missions: (1) discussing or studying the status of fishery resources, (2) formulating relevant investigation plans, (3) verifying protection measures related to fishery resources, (4) exchanging catch statistics, and (5) publishing data and documents.

Since the 1990s, international fisheries organizations have focused on strengthening fisheries management, playing an increasingly important role in the implementation of regulatory measures and international cooperation with law enforcement. Their priority has transferred from research and consulting to management and supervision, with management as the top priority. Regional fishery bodies (RFB) have transformed into regional fisheries management organizations (RFMO). Their main functions are (1) developing conservation and management measures, (2) setting total allowable catches and fishing quotas for member states, (3) promoting and regulating national fisheries management, (4) dealing with fishing issues, (5) implementing provisions of international law, (6) developing monitoring measures, and (7) approving of joint law enforcement and enforcement procedures, such as fishing vessel boarding, inspection, and so on.

2.10   Major Issues and Development Trends of Global Fisheries

2.10.1   Major Issues of Global Fisheries

2.10.1.1   Overfishing and Fishing Overcapacity

Currently, fishing capacity is not solely defined by the number of fishing vessels. With science and technology development, navigation, fish exploration, and fishing techniques are constantly improving, and the technologies are constantly updated. Therefore, overcapacity can be understood as the fishing capacity exceeding the regeneration capacity of fishery resources. This phenomenon has led to a decline in fishery resources, especially in economically important demersal fish. Excessive investment and/or government subsidies are likely some of the main reasons for fishing overcapacity. According to an FAO report in 2006, 31.4% of fish species were estimated as fished at a biologically unsustainable level and therefore overfished, 58.1% were fully fished, and 10.5% were underfished.

2.10.1.2   Bycatch and Discards

Bycatch refers to species that are accidentally caught or mixed in the process of catching targeted species. Bycatch is sometimes called nontarget species. The bycatch issue is related to the choice of fishing gear, which means the fewer the type and quantity of non-target species captured, the higher the selectivity of the fishing gear. According to a fishing gear analysis, the bycatch situation is as follows: shrimp trawling accounts for 62%, tuna longline fishing accounts for 29%, fixed nets account for 23%, and bottom trawling accounts for 10%. Discards are the catches that are thrown overboard. When the fish tank on a ship is full, the low-value species or species that are not welcome because of local customs are discarded. The former mainly refers to low-value species abandoned in marine capture to ensure the quota of economic species. As for the latter, for example, some ethnic groups do not eat fish with no scales. According to the FAO report, the waste from 1988 to 1990 was an estimated 17.9–39.5 million tons. After years of improvement, this waste was reduced from 6.9–8 million tons from 1992 to 2001. Both bycatch and discards lead to the decline and waste of fishery resources.

2.10.1.3   IUU Fishing at High Seas

Regional fisheries management organizations take measures to conserve high seas fishery resources, including restrictions on the number of fishing vessels and catches. Fishing vessels that are allowed to fish should report their location and catch as required. IUU fishing (illegal, unreported, and unregulated fishing) in the high seas can be defined as follows: (1) fishing activities conducted by national or foreign vessels in the high seas, without permission of the state or in contravention of its laws and regulations, or fishing activities conducted with permission of the state while in an illegal fashion; (2) fishing activities conducted by vessels that violate the conservation and management measures of regional fisheries organizations or the relevant provisions in international law, that fail to report according to prescribed procedures, that refuse to report, or that intentionally misreport the catches and position of the vessel; (3) fishing activities conducted by vessels that refuse to report or misreport the catches and position of the vessel to the state; (4) fishing activities conducted in waters under the jurisdiction of an organization by vessels of states who are not members of the regional fisheries management organization. Fishing activities defined as IUU fishing lead to a loss of control, overfishing capacity, and the decline of fishery resources in the high seas. The FAO adopted the *National Plan of Action on IUU fishing* in 2001, requiring countries and international fisheries organizations to take measures that regulate IUU fishing and conserve fishery resources.

2.10.1.4 Conservation of Ecosystems and Endangered Animals (Including Bycatch, Seabirds, Turtles, and Sharks)

Ecosystem protection prevents the growth and decline of one species from affecting the survival of other species in an aquatic ecosystem. By the 1990s, the conservation of fishery resources in the field of fisheries science had received much attention, but it was focused on adopting measures to prevent a particular species from decline, and such measures might include closing fishing areas, closing fishing seasons, restricting mesh size, and restricting the minimum length of fish, among others. In fact, interactions between species in ecosystems are very common. For example, the bycatch or accidental capture of seabirds, sea turtles, and sharks results in the severe deterioration or destruction of ecosystems in many marine and inland waters. Therefore, ecosystem-based conservation of fishery resources was proposed in the 1990s. The FAO adopted the *Reykjavik Declaration on Responsible Fisheries in the Marine Ecosystem* (referred to as the Reykjavik Declaration) at the Conference on Responsible Fisheries in the Marine Ecosystem held in Reykjavik, Iceland, from October 1st to 4th, 2002. The *Reykjavik Declaration* integrates ecosystems into fisheries management to ensure the effective conservation and sustainable use of an ecosystem and its biological resources.

2.10.1.5 Protection of the Aquatic Environment

Aquatic ecological and environmental problems not only involve aquatic environmental pollution, such as harmful algal blooms caused by pollution from land and ships, but also acid rain, El Niño, and La Niña caused by air pollution and global warming. This pollution directly affects the survival of aquatic organisms and the performance of fisheries production, and the situation will become increasingly serious. At the same time, many serious pollution problems also arise with aquaculture development. For example, fish and shrimp are used directly as feed, including the residue from fish drugs used to control disease, and the waste water discharge from farms also brings about pollution.

2.10.1.6   Quality and Safety of Fishery Products (Including Residues from Farmed Products)

Accidents concerning the quality and safety of fishery products occur occasionally due to pollution in aquaculture waters, feed or fish drugs containing additives, and anticorrosion measures forbidden although still used in processing, which directly cause serious negative impacts on human health. The quality and safety of fishery products have attracted the attention of the international community.

2.10.2   Development Trends of International Fisheries

2.10.2.1   Higher Priority on the Sustainable Development and Utilization of Fishery Resources and the Sustainable Development of Fisheries

Since the 1990s, the international community has been giving more priorities to the sustainable development and utilization of fishery resources and the sustainable development of fisheries. In 1992, the United Nations adopted *Agenda 21* at the Global Environment and Development Summit in Rio de Janeiro, Brazil. Agenda 21 has raised awareness on tackling the tension between environment and development, and it has indicated the global willingness of cooperation and high-level political commitment to sustainable development in the twenty-first century. Chapter 17 of Part 2 of Agenda 21 deals with issues such as the protection of oceans, closed seas, semi-enclosed seas, and coastal areas, as well as the protection, rational use, and exploitation of living marine resources. Moreover, this plan requires international government, intergovernmental, or nongovernmental organizations to work together to cope with these issues.

The FAO adopted the *Cancun International Responsible Fishing Declaration* (referred to as the *Cancun Declaration*) at the International Conference on Responsible Fishing held in Cancun, Mexico, from May 6th to 8th, 1992. The declaration states that "responsible fishing" refers to the sustainable utilization of fishery resources in harmony with the environment, the use of capture and aquaculture practices that are not harmful to ecosystems or resources, meeting the required sanitary standards, increasing the added value of fishery products, and providing consumers with good quality products at lower prices, among others. The FAO was also required to draft the *Code of Conduct for Responsible Fishing* (referred to as the

Code of Conduct) according to this declaration. Following discussions at FAO COFI, the *Code of Conduct for Responsible Fisheries* was adopted in 1995 and has become an international guideline for fisheries management. The *Code of Conduct* requires that states assume responsibility while engaging in capture, aquaculture, processing, transport, marketing, international trade, scientific research in fisheries, and so on. These responsibilities involve the following issues: (1) sustainably utilizing fishery resources in harmony with the environment; (2) using capture and aquaculture practices that are not harmful to ecosystems, resources, or their quality; (3) meeting the required sanitary standards and increasing the added value of fishery products; (4) providing consumers with good quality products at lower prices; and (5) supporting the development of fishery scientific research and other forms of research.

To protect marine ecology and prevent accidental capture of marine mammals, sea turtles, and sea birds, the UN General Assembly adopted the resolutions 44/225, 45/197, and 46/215 in the *Large-scale Pelagic Driftnet Fishing and Its Impact on the Living Marine Resources of the World's Oceans and Seas* in the 44th, 45th, and 46th sessions in 1989, 1990, and 1991, respectively. These resolutions stipulate that large-scale pelagic driftnet fishing in the open seas of oceans and seas, including in closed seas and semi-closed seas, would be fully forbidden from the date of January 1st, 1993.

Following the United Nations Environment and Development Summit held in Brazil in 1992, the United Nations held the World Summit on Sustainable Development in Johannesburg, South Africa, in August 2002 and adopted the *World Summit on Sustainable Development Plan of Implementation* (referred to as WSSDPOI). According to *Agenda 21*, the main measures for achieving sustainable fisheries are as follows: recovering fishery resources to maximum sustained production by 2015, implementing the *Code of Conduct for Responsible Fisheries* and its related plans, strengthening the management of international fisheries organizations, eliminating IUU fishing, and so on.

2.10.2.2 Projections of Global Fisheries Production

In 2002, the FAO coordinated a projection in which countries and organizations forecasted global fisheries productions in 2010, 2015, 2020, and 2030 (FAO 2002). These forecasts were based on FAO fishery statistics in recent years, the potential of different fisheries productions, and the increasing population trend, among other factors. Some forecasts were amended in *The State of World Fisheries and Aquaculture (SOFIA) 2006*. As shown in Table 2.20, there are significant differences among the results of SOFIA 2002, FAO fishery statistics, and research findings of IFPRI. Among them, SOFIA 2002 maintains that in 2010, 2020, and 2030, there will be no obvious increase in marine fisheries or inland fisheries production and that production will stabilize at 93 million tons. The FAO and IFPRI do not distinguish between marine fisheries and inland fisheries, but they both believe that there will be a further increase in fisheries production, reaching 105–116 million tons. As for the

Table 2.20 Projections of global fisheries production in 2010, 2015, 2020, and 2030 (unit: million tons) (FAO 2006)

	2000	2004	2010	2015	2020	2020	2030
	FAO statistics	FAO statistics	SOFIA 2002	FAO research[a]	SOFIA 2002	IFPRI research[b]	SOFIA 2002
Marine capture	86.6	85.8	86		87		87
Inland capture	8.8	9.2	6		6		6
Capture subtotal	95.4	95.0	92	105	93	116	93
Aquaculture	35.5	45.5	53	74	70	54	83
Total production	131.1	140.5	146	179	163	170	176
For food purposes	96.9	105.6	120		138	130	150
Percentage for food purposes (%)	74	75	82		85	76	85
Not for food purposes	34.2	34.8	26		26	40	26

Note: [a]FAO (2004); [b]International Food Policy Research Institute (2003)

development trends of aquaculture production, SOFIA 2002 and the FAO hold a similar forecast that there will be major growth, with an aquaculture production estimated by the FAO at 74 million tons in 2015 and by SOFIA 2002 at 53 million tons, 70 million tons, and 83 million tons in 2010, 2020, and 2030, respectively. However, IFPRI believes that there will only be a slight increase in 2020 at 54 million tons. These forecasts can serve as a reference for our study of related fisheries issues.

2.10.2.3   Development Trends of Global Fisheries

In light of the current status and projections of global fisheries, the development trends of world fisheries are mainly illustrated as follows.

Priority Shift from Marine Fishery Resource Exploitation to Fisheries Management and Promotion of Regional Management of Marine Capture

Since the 1990s, the international community has made it clear that there is a priority shift to management from the exploitation and utilization of marine fishery resources. Regional management of marine capture has also been increasingly promoted. Most of the regional fisheries organizations (referred to as RFO) have been transformed into regional fisheries management organizations (referred to as RFMO), which play a unique role in promoting the conservation and management of fishery resources and international cooperation. Organizations cover the management of marine fisheries in all oceans. After the adoption of the UNFSA in 1995, the South East Atlantic Fisheries Organization (SEAFO) and the Western and Central Pacific Fisheries Commission (WCPFC) were established. In 2004, the FAO

Council decided to establish the Southwest Indian Ocean Fisheries Commission (SWIOFC). The FAO also clarified that, in the future, only by developing fisheries associations and strengthening fisheries management will it be possible to further develop fisheries production and improve efficiency.

Further Aquaculture Development as the Primary Means for Future Fisheries Development

China enforces a fisheries policy that is "aquaculture-oriented; aquaculture-capture-processing-coordinated; adapted to local conditions; and prioritizes local advantages." At present, marine and freshwater aquaculture production in China exceeds capture production and accounts for 70% of the global aquaculture production, which greatly promotes adjustment of the global fisheries industry structure. Aquaculture development has received widespread attention. Norway has long been a typical marine capture country and has become a major aquaculture country for Atlantic salmon and also a driving force for fisheries development in other countries. In a fisheries report in 2006, the FAO maintained that 50% of the global fishery products for human consumption originated from aquaculture. Projections of the global fisheries productions in 2015, 2020, and 2030 also show that further aquaculture development is the primary means for future fisheries development. With the development and improvement of storm-resistant marine cage culture, marine aquaculture may be the primary contributor to aquaculture production.

Higher Priority on Ecological Fisheries

Both capture and aquaculture must be carried out and developed in accordance with the requirements of ecological fisheries. The *Reykjavik Declaration* clearly states that a code of conduct for responsible fisheries should be developed to integrate ecosystems into fisheries management and promote the development of ecological fisheries to ensure effective conservation and sustainable use of ecosystems and their biological resources.

Development of Ornamental Fish Aquaculture

Ornamental fish aquaculture is the aquaculture of fish not destined for human consumption as food, and it has been recognized as a potential global industry. In addition to seawater and freshwater ornamental fish, live aquatic animals are also targets for ornamental fish aquaculture. Countries have taken active measures to boost their ornamental fish aquaculture and trade, with a view to increase rural employment and income and to enhance foreign exchange earnings. However, attention should be paid to the prevention and control of fish disease in this process of development. It would be disastrous if a disease were to spread worldwide.

Ecotourism Development

Ecotourism is a newly emerging industry that is likely to be developed and popularized in different countries. Generally speaking, most countries are striving to promote aquaculture-related ecotourism. Utilizing cage culture and ponds in lakes and reservoirs for ecotourism are popular in Russia, Ukraine, Belarus, and Moldova in Central and Eastern Europe and in some countries around the Baltic Sea. Some marine countries have developed ecotourism, based on large, storm-resistant cages or platform cages in the open seas, as well as coral reef exploration combined with diving, among others.

Development of Deep Processing of Fishery Products and Improving the Added Value of Fishery Products

With marine animals, plants, and microbes as raw materials and modern techniques, such as separation, purification, structural identification, optimization, and evaluation of pharmacological effects, as the processing means, medicines have been developed using raw materials with clear pharmacological activities, and processing waste has been recycled. The following technologies are currently rapidly developing throughout the world: developing medicines by using viscera from farmed Atlantic salmon, making leather products using farmed tilapia skin, producing chitosan using crab and shrimp shells, and producing antiarthritis medicines using green mussels. These technologies turn waste into treasure, reducing pollution and increasing the added value of aquatic products.

References

FAO (2002) The State of World Fisheries and Aquaculture 2002. Rome
FAO (2004) Future prospects for fish and fishery products: medium-term projections to the years 2010 and 2015. FAO Fisheries Circular FIDI/972–1. Rome
FAO (2006) The State of World Fisheries and Aquaculture 2006. Rome
FAO (2016) The State of World Fisheries and Aquaculture 2016. Rome
International Food Policy Research Institute (2003) Fish to 2020: supply and demand in changing global markets, by C. Delgado, N. Wada, M. Rosegrant, S. Meijer and M. Ahmed. International Food Policy Research Institute, Washington, DC
Lehner B, Doell P (2004) Development and validation of a global database of lakwes, reservoirs and wetlands. J Hydrol 296(1–4):1–22

Chapter 3
Overview of Major Fishery Countries in the World

Jiahua Le and Xinjun Chen

3.1 Brief Introduction

The fisheries industry, on the whole, is a primary industry, or "primary production industry." However, the capture industry in the fisheries industry, especially commercial fishing, has industrial properties; the large-scale industrialized aquaculture industry not only has industrial properties, but some have been categorized as members of the high and new technology industry. The aquatic products processing industry is a secondary industry. Therefore, there are significant differences in the attribution of fisheries in the national economic sectors of the countries concerned: with fisheries established independently in some countries, such as Norway's Ministry of Fisheries; with fisheries established together with agriculture, forestry, and fisheries, such as the Ministry of Agriculture, Forestry, and Fisheries of Japan; and with those jointly established for fisheries and oceans, such as the Ministry of Maritime Affairs and Fisheries of South Korea. According to the division of China's national economic sectors, fisheries fall under the agriculture sector.

According to the FAO fishery statistics, the fisheries productions of China, Japan, the USA, Indonesia, and India have ranked among the highest in the world since 2010. Among them, China is a superpower regarding capture and aquaculture. From 2010 to 2016, China's fisheries production (including aquatic plants) ranged from 63.4 to 81.53 million tons, with an average annual production of 72.9 million tons,

J. Le
College of Economics and Management, Shanghai Ocean University, Lingang New City, Shanghai, China
e-mail: jhle@shou.edu.cn

X. Chen (✉)
College of Marine Sciences, Shanghai Ocean University, Lingang New City, Shanghai, China
e-mail: xjchen@shou.edu.cn

© Science Press & Springer Nature Singapore Pte Ltd. 2020
X. Chen, Y. Zhou (eds.), *Brief Introduction to Fisheries*,
https://doi.org/10.1007/978-981-15-3336-5_3

accounting for 37–40% of the global fisheries production and showing a continued growth trend during this period; Indonesia's fisheries production (including aquatic plants) ranged from 11.6 to 23.2 million tons, with an average annual production of 18.1 million tons, accounting for 6.9–11.5% of the global fisheries production and showing a trend of sustained growth during this period; India's fisheries production (including aquatic plants) ranged from 8 to 10.8 million tons, with an average annual production of 9.38 million tons, accounting for 4.5–5.3% of the global fisheries production and presenting a trend of sustained slight growth during this period; the USA's fisheries production (including aquatic plants) ranged from 5.2 to 6.1 million tons, with an average annual production of 5.82 million tons, accounting for 2.9–3.3% of the global fisheries production and presenting a trend of sustained stability during this period; Japan's fisheries production (including aquatic plants) ranged from 4.3 to 5.4 million tons, with an average annual production of 4.78 million tons, accounting for 2.1–3.2% of the global fisheries production and showing a continuous slight downward trend during this period.

3.2 Fisheries in China

3.2.1 Natural Environment and Conditions of Fisheries

3.2.1.1 Natural Environment of Marine Fisheries

The Chinese mainland faces the Bohai Sea, the Yellow Sea, the East China Sea, and the South China Sea to the east and south, which are marginal seas of the Pacific Ocean. The mainland coastline from the Yalu River Estuary to the Beilun Estuary covers a total length of more than 18,000 km. There are more than 6500 islands, each with an area of more than 500 m^2, and these island coastlines total more than 14,000 km. The annual runoff from coastal rivers into the sea is approximately 1.5 trillion m^3. These four major sea areas cover an area of 4.827 million km^2, and the continental shelf covers an area of 1.48 million km^2.

There are four sea areas in China mainland as follows:

(1) The Bohai Sea, China's enclosed sea. This area is bounded by the connection between the western corner of Laotie Mountain and the islands of Jingmiaodao and Penglaijiao. The western boundary is formed by the Bohai Sea and the eastern boundary the Yellow Sea. The Bohai Sea covers an area of 77,000 km^2, half of which has a water depth of 20 m or less and is only 85 m deep in the Laotie Mountain waterway. The characteristics of its sediment distribution are that the sediment particles around the coast are relatively fine and that fine sand is widely distributed and gradually coarsens toward the central basin.

(2) The Yellow Sea, which is connected to the Bohai Sea. The southern boundary is bounded by the connection between the northern corner of the Yangtze River estuary and the southwestern corner of Jeju Island in South Korea, which is connected to the East China Sea, covering an area of approximately 380,000 km^2.

The Yellow Sea is divided into three parts: the north, the middle, and the south, that is, the northern Yellow Sea is to the north of the connecting line between Chengshan Cape in Shandong Province and Changshan in North Korea, the central Yellow Sea is to the north of 34°N, and the southern Yellow Sea is to the south of 34°N. The whole Yellow Sea area is located on the continental shelf, with only one Yellow Sea trough in the southeastern part of the northern Yellow Sea. The sand gullies along the northern coast of Jiangsu are radially distributed in depth, and the bottom sediment is mainly composed of fine sand.

(3) The East China Sea. The northern East China Sea is connected to the Yellow Sea, the northeastern East China Sea is connected to the Sea of Japan through Tsushima Strait, and the southern East China Sea is connected to the South China Sea by the connection between Zhaoan County in Fujian Province and Eluanbi in Chinese Taipei. The East China Sea is the largest and widest fan-shaped sea area in the four major sea areas, and the total area is approximately 770,000 km^2, of which the continental shelf area accounts for approximately 74%, that is, more than 570,000 km^2. Most of the water depth is 60–140 m, the water depth of the outer edge of the continental shelf is 140–180 m, and the deepest part of the Okinawa Trough is 2719 m. The average water depth of Taiwan Strait is 60 m, and the Taiwan Strait beach in the southwestern Penghu Islands is 30–40 m, and the shallowest depth is 12 m, constituting a sill, which has some impact on the water exchange between the East China Sea and the South China Sea. The bottom sediment of the East China Sea continental shelf is generally bounded by a water depth of 60–70 m, with terrigenous deposits in the west, mainly composed of ooze, muddy sand, and silt, and with coastal shallow sea deposits, mainly composed of fine sand. The open area at the southern end of Taiwan Strait is composed of fine sand, medium-coarse sand, and fine-to-medium sand, with gravel in the inshore and volcanic materials, gravel, and bedrock in the southwestern Penghu Islands.

(4) The South China Sea. The South China Sea covers an area of 3.5 million km^2, and the northern continental shelf covers an area of 37.4 km^2. The terrain is shallow in the surrounding area and deep-set in the middle, showing the shape of a deep-sea basin. In addition to the Nansha Islands and other islands and reefs in the south, there is also the famous Sunda Shelf. The deepest part of the Beibu Gulf is 80 m, most of which is 20–50 m deep, and the sea floor is flat. As for the sediment distribution, the northern continental shelf is similar to that in the East China Sea, and the eastern side of Beibu Gulf is dominated by clay and ooze, surrounded by coarse sand and gravel.

The abovementioned four sea areas are semi-enclosed seas, and their marine hydrology is significantly different from that of oceanic seas, depending on the influence of weather and water systems on the mainland and its offshore currents, including tidal currents, among others. The distribution of the isotherms and isohalines in each water layer is generally high in the outside and low in the inside and high in the south and low in the north, with obvious seasonal variations. The ocean circulation mainly consists of two systems: the alongshore current and the Kuroshio Current in the Chinese mainland. The former generally flows from north to south along the coast, and the Kuroshio Current flows to the northeast all year, except for the South China Sea branch. The movement and migration of fishery resources are directly related to the influence of mutual fluctuation and topography. Since the entry into force of the UNCLOS, in the abovementioned four sea areas, except the Bohai Sea – China's enclosed sea, that is, in areas of the Yellow Sea, the East China Sea, and the South China Sea, China has disputed the demarcation of the 200-nautical-mile exclusive economic zone or continental shelf with neighboring countries.

3.2.1.2 Natural Environment of Inland Fisheries

China boasts vast inland waters, with a total area of approximately 0.176 million km^2, occupying 1.84% of the total land area, including more than 66.67 thousand km^2 of rivers, more than 73.34 thousand km^2 of lakes, more than 20 thousand km^2 of reservoirs, and more than 13.33 thousand km^2 of ponds. There are 104 rivers flowing more than 300 km, of which 22 rivers extend over 1000 km. There are more than 2800 lakes with an area of over 1 km^2 and 18 lakes with an area of over 666.7 km^2, mainly including Poyang Lake, Dongting Lake, Taihu Lake, Hongze Lake, Chaohu Lake, Hulun Lake, Nam Lake, Xingkai Lake, South West Lake, and Bosten Lake, among others. There are 87,000 reservoirs, including 328 large-sized, 2333 medium-sized, and 84,000 small-sized reservoirs. China is a country with vast territory, spanning 49 latitudes, 62 longitudes, and 5 climatic zones. Therefore, there are huge differences in natural geographic, climate, and hydrological conditions, which make fishery resources in inland waters complex and diversified.

The main inland waters in China mainland can be broadly divided into the following categories:

(1) Heilongjiang River and Liaohe River Basin, which are located in the cold temperate zone, with uneven precipitation distribution. The freezing period is from November to April of the following year, and the navigation period is from May to November. The average monthly water temperature is 7–24 °C, and the water surface is mostly distributed in the north, with different ice periods. The water area in this river basin accounts for approximately 9% of the country.

(2) The Yellow River and Haihe River Basin, which are located in the southern temperate zone, with insufficient precipitation and long periods of sun and an annual average monthly water temperature of 3–28 °C. The water surface is mainly distributed in the flat source area of the lower reaches of the river, and the water area of the whole basin accounts for approximately 11% of the country.

(3) The Yangtze River Basin, which is located between the southern temperate zone and the subtropical zone, with abundant precipitation, mild climate, and an annual average monthly water temperature of 6–29 °C. The plain areas in the middle and lower reaches of the basin are crisscrossed by rivers and streams, with a high density of lakes and interconnected water networks, and these water areas account for approximately 46% of the main river basins in China.

(4) The Pearl River Basin, which is located in the subtropical zone, with especially abundant precipitation and an annual average monthly water temperature of

13–30 °C. These water areas of the whole basin account for approximately 7% of China.

(5) The regions of Xinjiang, Qinghai, and Tibet, which are located in the plateau area. This is the largest plateau lake group distribution area not only in China but also in the world, and its water area accounts for 25% of China. There are also landlocked slightly alkaline surface waters, such as Qinghai Lake.

According to the natural conditions in the abovementioned major river basins, the Yellow River Basin, the Yangtze River Basin, and the Pearl River Basin are more suitable for aquaculture, especially the middle and lower reaches of the Yangtze River Basin, where there are extensive interconnected water networks, and these water areas account for almost half of the total water areas in China. However, Xinjiang and Qinghai-Tibet Plateau, with a large number of plateau lakes, have unique favorable conditions for developing cold-water fish farming.

3.2.2 Fish Species

3.2.2.1 Marine Fishery Resources

At present, China has a wide variety of marine fishery resources. There are more than 2000 kinds of marine fish, approximately 40 kinds of marine animals, approximately 80 kinds of cephalopods, 300 kinds of shrimp, more than 800 kinds of crabs, approximately 3000 kinds of shellfish, and 1000 kinds of marine algae. Additionally, there are approximately 200 kinds of fish recorded in the fishery statistics and market circulation sales.

According to the water depth that fish inhabit, the scope of the sea areas where fish are distributed, and the characteristics of relevant fishery resources, fishery resources can be classified as follows:

(1) The pelagic species, the fishery resources mainly inhabiting the middle and upper water layers, such as mackerel, Spanish mackerel, and so on

(2) The demersal or near-demersal species, fishery resources that mainly inhabit the bottom or near-bottom waters, such as small yellow croaker, large yellow croaker, hairtail, anglerfish, codfish, and so on

(3) The estuarine fishery resources, which mainly inhabit river estuaries, such as mullet, pike, sea bass, migratory tapertail anchovy and long-tailed anchovy, and so on

(4) The highly migratory species, which include fishery resources that generally grow up in the ocean and regularly migrate over long distances, such as tuna, bonito, marlin, swordfish, and so on

(5) The anadromous migratory species, which include fishery resources that grow up in the ocean and return to the original spawning river to breed, such as salmon

(6) The catadromous migration species, which include the fishery resources that, contrary to anadromous migratory fish, grow up in the river and return to the ocean to breed, such as eel

The statistics on the exploitation and utilization of marine fishery resources in China shows that there are no super fishing targets and that the actual annual catch of anchovy alone once exceeded 1 million tons. There are more than 40 species with an annual catch of more than 10,000 tons, mainly including hairtail (*Trichiurus japonicus*), filefish (*Thamnaconus modestus*), Japanese scad (*Decapterus maruadsi*), large yellow croaker (*Pseudosciaena crocea*), Japanese mackerel (*Scomber japonicus*), Pacific herring (*Clupea pallassi*), silvery pomfret (*Pampus argenteus*), Spanish mackerel (*Scomberomorus niphonius*), greater lizardfish (*Saurida tumbil*), bigeye snapper (*Priacanthus tayenus*), spinyhead croaker (*Collichthys lucidus*), *Loligo japonica*, *Loligo chinensis*, and *Penaeus chinensis*, among others. Due to resource fluctuation, there are more than 30 fish species with catches exceeding 10,000 tons, and there are indications that there have been more catches of crustacean resources in recent years.

3.2.2.2 Inland Fishery Resources

China's inland waters are rich in fishery resources. According to a survey, there are more than 800 fish species in China, of which over 760 species (including subspecies) are purely freshwater fish and over 60 species are migratory fish. In recent years, more than ten fish species have been introduced and transplanted from abroad.

Within the distribution of fish resources in inland watersheds or regions, there are 381 species in the Pearl River Basin, 370 species in the Yangtze River Basin, 191 species in the Yellow River Basin, 175 species in the Heilongjiang River Basin, more than 50 species in Xinjiang, and 44 species in Tibet in the western plateau. On the whole, the species distribution tends to decrease from east to west and from north to south.

In terms of fingerlings, the proportion of Cyprinidae fish is the highest, with an average of 50–60% in all water systems nationwide. There are also a considerable number of migratory fish in estuary areas, including anadromous and catadromous fish. In addition to fish, there are plenty of shrimp and crabs, such as *Macrobrachium*, *Palaemonidae*, and Chinese mitten crab (mitten crab), snails, and mussels – for example, *Hyriopsis cumingii* and wrinkles mussels are both mother mussels for raising pearls in freshwater. Some are listed as national-level rare and protected animals, such as *Lipotes vexillifer* (Chinese river dolphin), finless porpoise (cowfish), *Acipenser sinensis* (Chinese sturgeon), paddlefish, Chinese alligator, and *Palea steindachneri* (wattle-necked softshell turtle), among others. Some are listed as endangered species, such as *Trachidermus fasciatus* and *Schizothorax taliensis*.

3.2.3   Current Status of Fisheries Development

China's fisheries have played an important role in the historical development of global fisheries. China's aquaculture, especially its freshwater aquaculture, has a positive impact on global aquaculture. China's marine fishing technologies, such as trawling, purse seining, and trapping fish with lights, have long been spread abroad and have promoted the development of global fisheries technology. In recent years, with the rapid development of China's fisheries, aquatic products production, aquaculture production, the import and export trade of aquatic products, the number of fishing vessels, the total number of fishermen and fishery workers, and deep-sea fisheries production have ranked first in the world for many years, playing an increasingly important role in world food security. Meanwhile, China's status and role in global fisheries has also been increasingly enhanced. China's aquatic products production accounted for less than 5% of the global total in 1950, approximately 40.4% in 2016, and has stabilized at approximately 38%.

According to the world fisheries statistics released by the FAO and China's Fisheries Statistical Yearbook, the global capture and aquaculture production in 2016 was approximately 171 million tons (fresh weight), including fish, crustacean, mollusk, other edible aquatic animals (such as sea cucumber, jellyfish, Echinus, and sea squirt), and algae. China's total fisheries production in 2016 was approximately 69.01 million tons (fresh weight), accounting for 40.4% of the global total. Aquaculture has played a significant global role, with marine culture and freshwater aquaculture accounting for 68.4% and 61.8% of the global total, respectively, while marine and freshwater capture productions were relatively low, at 19.3% and 19.9%, respectively.

According to the structure of the fisheries industry in China and around the world, China's fisheries production is concentrated in aquaculture, and its marine and freshwater aquaculture production accounts for 74.5% of the total production, while capture only accounts for 25.5% of the total production. Capture and aquaculture account for 53.2% and 46.8% of the global fisheries production, respectively (Figs. 3.1 and 3.2). The proportion of global marine capture is as high as 46.4% of the total production, while China's marine capture only accounts for 22.1% of the total production. Therefore, the fishery industry structure in China is focused on aquaculture, which has enhanced China's role in protecting global marine fishery resources. In terms of the capture industry structure, the number of marine fishing workers in China was approximately 1.002 million in 2016, with approximately 359,000 fishing vessels (including some unlanded fishing vessels),

Fig. 3.1 Fisheries production structure of China mainland in 2016 (FAO 2018)

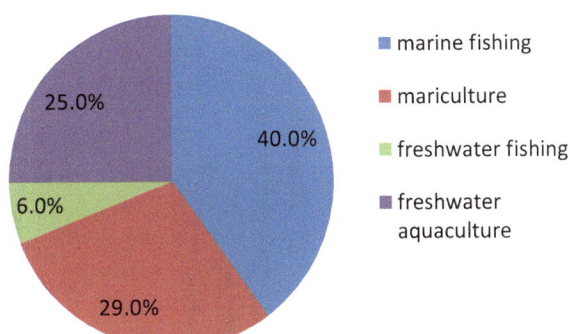

Fig. 3.2 Global fisheries
production structure in 2016
(FAO 2018)

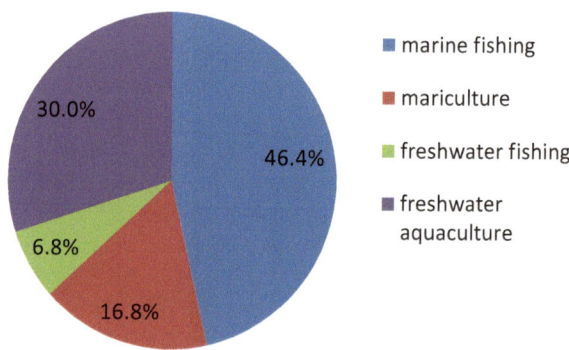

15.271 million tons of fisheries production, and 15.2 tons of per capita production of
fishermen. The productivity remained at a low-to-intermediate level, so it would be
necessary to further improve the rational and scientific utilization of fishery
resources and to effectively improve the level of marine capture productivity on
the premise of protecting the marine ecological environment and the livelihoods of
coastal fishermen.

In terms of marine fisheries, China's marine capture production in 2016 was
approximately 15.27 million tons, accounting for 18% of the global production, of
which inshore capture production was 13.283 million tons and deep-sea capture
production was 1.98 million tons. In 2016, global marine aquaculture production
reached 28.7 million tons, and China's marine aquaculture production reached
19.63 million tons (16.37 million tons of algae and shellfish products, accounting
for 83.4% of the global marine aquaculture production), comprising 68.4% of the
global marine aquaculture production.

In terms of inland fisheries, the global total freshwater capture production in 2016
was 11.6 million tons, and China's freshwater capture production was 2.318 million tons, accounting for 19% of the total. In 2016, China's freshwater aquaculture
production reached 31.79 million tons, accounting for 61.8% of the global total
production, while the total aquaculture production of the 20 countries ranked from
second to twentieth only accounted for approximately 35% of the global total
production, highlighting that China's freshwater aquaculture plays an important
role in the development of global freshwater aquaculture. Developing countries
occupy an overwhelming position in freshwater aquaculture, basically occupying
the 20 highest rankings.

Since the 1990s, China has developed into a major aquaculture country
(Table 3.1). In 2016, the ratio of capture to aquaculture reached 25.50:74.50,
while the vast majority of aquaculture production was from freshwater aquaculture.
Therefore, in this sense, aquaculture is "the aquaculture of developing countries."
The reason is that freshwater aquaculture species are mainly low-value aquatic
products, which meet the basic protein needs of developing countries and less
developed countries. As for the classification of freshwater aquaculture products,
China ranks first in the production of fish, crustaceans, shellfish, algae, and other

Table 3.1 Ratio of capture production to aquaculture production in China mainland from 1990 to 2016 (Chen and Zhou 2018)	Year	Capture:Aquaculture
	1990	45.70:54.30
	2000	34.20:65.80
	2005	33.00:67.00
	2010	28.70:71.30
	2015	26.30:73.70
	2016	25.50:74.50

animals. This production of crustaceans, shellfish, algae, and other animals accounts for more than 90% of the global total. Shrimp and crab are traditional breeding species in China and are popular among the Chinese people. During the decade from 1990 to 2000, the proportion of capture production in total aquatic products production decreased by 11.5%, ranking second among China, Indonesia, India, the USA, and Japan. Additionally, this figure fell by 8.7%, ranking third among the top five countries during the 16 years from 2000 to 2016. The results show that China's aquaculture has been developing continuously since the 1990s. The aquaculture production has increased by eight times compared with 1990, with an average annual growth rate of 8.3%. It has become an important mission for China's aquaculture to improve product quality, reduce aquaculture farms, increase unit production, and produce species catering to market demand both at home and abroad.

3.2.4 Development Trends of Fisheries in China

China has a long fisheries history, enjoying vast sea areas, widely distributed inland waters, and abundant fishery resources, which provide favorable conditions for fisheries development. Thanks to the reform and opening up, China has made rapid progress in fisheries, with the annual production ranking first in the world for decades. However, China is not yet a strong nation in the capture industry, and there is still much to be researched and developed in fisheries science, production technology, and fisheries management.

In the future, China will firmly stick to the philosophy of fisheries development as being innovative, coordinated, green, open, and shared, with the aims of improving quality and efficiency, reducing production while increasing income, achieving green development, and enhancing fishermen income, by developing healthy aquaculture, fishing moderately, conserving resources, and strengthening industries. The priority should be given to promoting the supply-side structural reform of fisheries, accelerating the transformation of fisheries development methods, improving standard levels, greening, industrializing, organizing, and sustaining development, thus improving the quality benefits and competitiveness of fisheries development, and realizing fisheries modernization with Chinese characteristics – high production, safe products, resource conservation, and being environmentally friendly. Domestic capture production is expected to achieve "negative growth" and domestic marine capture production will be controlled within 10 million tons. The key missions are listed as follows: to vigorously conserve aquatic biological resources; to transform and upgrade the aquaculture industry; to reduce and control fishing efforts for the capture industry; to develop deep-sea capture in a standard and orderly way; to promote construction of the fishing port economic zone; to enhance integration of primary, secondary, and tertiary industries; and to improve the level of fisheries safety development.

3.3   Fisheries in Japan

3.3.1   Natural Environment and Conditions of Fisheries

3.3.1.1   Natural Environment of Marine Fisheries

To the east and south of Japan lies the Pacific Ocean, and Japan is bordered by the Sea of Japan and the East China Sea in the west and by the Sea of Okhotsk in the north. Japan is made up of Hokkaido, Honshu, Shikoku, and Kyushu, the four major islands and another 6847 small islands, known as the "Thousand Islands Country," of which 422 are inhabited. Japan has a coastline of 33,889 km, 2866 various types of fishing ports, 3 million km^2 of continental shelf area, and 4.47 million km^2 of sea area, making Japan sixth in the world in terms of the size of sea areas.

There are three sea areas in Japan, and they are illustrated as follows: (1) The Sea of Japan, which is located in the western part of the Japanese mainland, is the largest marginal sea in the northwest Pacific Ocean. The eastern boundary starts from the north at Sakhalin Island, the western boundary is Russia's Far East region, and the southern boundary is the Korean peninsula. The Sea of Japan covers an area of approximately 1.52 million km^2 and is 2300 km long from north to south and 1300 km wide from east to west, with a maximum water depth of approximately 3800 m and an average water depth of 1350 m. In addition to the terrestrial sediments, such as mud, sand, gravel, and rock debris in the littoral zone, the bottom sediments are mainly marine ooze sediments. (2) The Sea of Okhotsk, a marginal sea in the northwest Pacific Ocean, is 2460 km long from north to south and 1480 km wide from east to west, covering an area of approximately 1,528,000 km^2, with a maximum depth of 3658 m in the east, 1000–1600 m in the center and an average depth of 838 m. The total length of the coastline is 10,460 km. The bottom sediments are coarse gravel, fine gravel, and sand in the littoral zone and silty mudstone, silty clay, and argillaceous mud in the deep-sea areas. The Kuril Islands area is rich in volcanic clastic materials, which can form tuff sedimentary layers of various granularity. (3) The eastern side of the Pacific Ocean, which is located in the subtropical region, is mild and humid all year round, with more typhoons in summer and autumn, more plum rains in June, and more typhoons from August to September. The average annual precipitation is 2000 mm, and the average water depth is 188 m. The bottom sediments are covered with glauconite and biogenic deposits (including calcareous ooze), are widely distributed with abyssal clay containing diatoms, and are scattered with volcanic clastic materials.

The eastern side of the Pacific Ocean is surrounded by the warm Japanese Current (Kuroshio Current) flowing from north to south, and the cold Thousand Island Current (Oyashio Current) is formed in the northeast. On the western side of the Sea of Japan are the Tsushima Warm Current and Riemann Cold Current, forming a natural fishing ground at the intersection of the cold and warm currents, where there are abundant aquatic resources. The Hokkaido fishing ground used to be one of the three most famous fishing grounds in the world. There are disputes on the demarcation of the 200-nautical-mile exclusive economic zone and continental shelf with China and South Korea.

3.3.1.2 Natural Environment of Inland Fisheries

A large number of rivers and lakes are extensively distributed in Japan, covering an area of approximately 70,000 km^2. Among them, there are more than 2300 large and small lakes, artificial lakes, and reservoirs, with an area of approximately 300,000 km^2. The lakes with a water surface area of more than 100 km^2 include Lake Biwa, Lake Kasumigaura, Lake Salome (Yuanjian), and Lake Inawashiro, among others. The lakes are characterized by a small size, short flow distance, large drop, and significant variation of water carrying capacity due to seasonal variation. The total length of the rivers is approximately 320,000 km, of which there are 44 first-class rivers with basin areas of more than 1500 km^2, mainly including the Tone River, Ishikari River, Shinano River, and Kitakami River, among others.

The inland water surface of Japan mainly involves the following waters: (1) Shinano River, the longest river in Japan, is 367 km in length and flows through most parts of Nagano and Niigata. Its upper reaches are characterized by the most typical inland climate, while the northern part is affected by the climate in the north, showing complex climatic conditions. The middle and lower reaches show the climate characteristics of the Sea of Japan. From November to February of the following year, snowfall accounts for 40–50% of the annual precipitation, of which the mountainous area in the lower reaches of Shinano River receives approximately 3000 mm, and the mountainous area in the middle reaches approximately 2000–2500 mm. The southern part shows obvious climatic characteristics of the East China Sea. During the plum rain season from June to July each year, the annual precipitation of the flat area along the coast is at least 1900 mm, and the flat area near the mountains is approximately 2600 mm. (2) Lake Biwa, the largest lake in Japan, covers an area of 674 km^2, with a lakeshore length of 241 km and an average water depth of 41.2 m. Inland waters can not only be used as food resources but also as a destination for domestic leisure vacations. Aquaculture plays an important role in fisheries sustainability.

3.3.2 Fish Species

3.3.2.1 Marine Fishery Resources

Japan has a large variety of living marine resources. There are approximately 300 families and more than 3200 fish species and approximately 600 fish species with economic value. Additionally, there are approximately 150 fish species recorded in the fishery statistics and market sales.

According to the water layers that fish inhabit, scope of the sea areas where fish are distributed, and characteristics of relevant fishery resources, the fishery resources can be classified as follows: (1) The pelagic species mainly include mackerels, horse mackerel, sardine, anchovy, herring, sea bream, squid, and so on. (2) The demersal species mainly include cod, grouper, scallop, and so on. (3) The mammals mainly include white whale, sperm whale, blue whale, and so on. Moreover, there are also sea lion, shrimp, crab, tuna, skipjack, kelp, and so on.

According to statistics on the exploitation and utilization of marine fishery resources in Japan, the actual annual catch of mackerel alone is between 500,000 and 600,000 tons. There are approximately 10 species with annual catches ranging from 100,000 to 400,000 tons, mainly including skipjack, tuna, Japanese anchovy, salmon and trout, saury, horse mackerel, pollock, Atka mackerel, scallops, and so on. Furthermore, there are approximately 5 species with annual catches ranging from 10,000 to 100,000 tons, mainly including flounder fish, Japanese amberjack, Pacific sardine, snapper, swordfish, and so on. Due to resource fluctuation, there have been other fish resources and species captured in recent years.

3.3.2.2 Inland Fishery Resources

Japan is rich in inland fishery resources, with more than 1000 species of animals and plants, including approximately 46 fish species, 40 shellfish species, and 70 aquatic plants species. Its freshwater fish production accounts for more than 50% of the country's total, and pearl production has long enjoyed a good reputation.

In terms of fish species, the main fingerlings exploited are salmon and trout, catfish, and clam, and the main fingerlings cultured are trout, catfish, carp, eel, and so on. Cold-water fish, such as salmon and trout and catfish, are mainly distributed in mountainous areas where there are abundant water resources.

3.3.3 Current Status of Fisheries Development

Japan is one of the most developed countries with regard to fisheries in the world. In 2016, there were 85,000 fishery operating entities in Japan, more than 94% of which accounted for those involved in coastal fisheries, while small- and medium-sized fisheries (inshore fisheries) and large-scale fisheries (deep-sea fisheries) comprised a small proportion. Among the major fishing methods in coastal fisheries, hook fisheries occupied the largest proportion, followed by gillnet fisheries, shellfish-collecting fisheries, other fisheries, small trawling fisheries, algae-collecting fisheries, and so on. Inshore fisheries can be divided into inshore trawling fisheries, coiling net fisheries, stick-held dip net fisheries for saury, squid jigging fisheries, and trawling fisheries. However, there are many differences in the operation scale, which varies according to the variation of sea areas, and diversified fishing methods are generally adopted. At present, the fishing methods of deep-sea fisheries mainly include pelagic trawling, western bottom trawling, large- and medium-sized pelagic flow netting for skipjack and tuna, longline and gill netting in the northern ocean, the tuna longline fishery, pelagic skipjack javelin fishery, pelagic squid jigging fishery, and so on, and the production scale of the fisheries mentioned above is in a stable state.

The overall fisheries production of Japan in 1984 exceed 12.8 million tons, accounting for approximately 15% of the global total. However, since 1995, fisheries production has been rapidly reduced under the fluctuation of Japanese sardine (*Sardinops melanostictus*) resources. In recent years, Japan's fisheries production has kept declining due to the combined factors from resources, environment, labor force, production cost, and income. Japan's total fisheries production in 2016 was 4.359 million tons, a decrease of 5.9% compared with the previous year, accounting for 2.5% of the world total, the lowest recorded since 1995 when fisheries production statistics were first released in Japan, and this trend is likely to continue.

As for the fisheries production structure of Japan, most fisheries production involves marine fisheries. The production of marine capture and aquaculture is 4.296 million tons, accounting for 98.6% of the total production (Fig. 3.3), which exceeds the proportion of global marine fisheries in global fisheries production by 35.4%. This statistic shows that Japanese fisheries production is basically dependent on marine fisheries and that priority is given to marine capture, which accounts for 74.9% of the total fisheries production, 28.5% higher than the proportion of global marine capture in global fisheries production. In terms of the capture production

Fig. 3.3 Fisheries production structure of Japan in 2016 (FAO 2018)

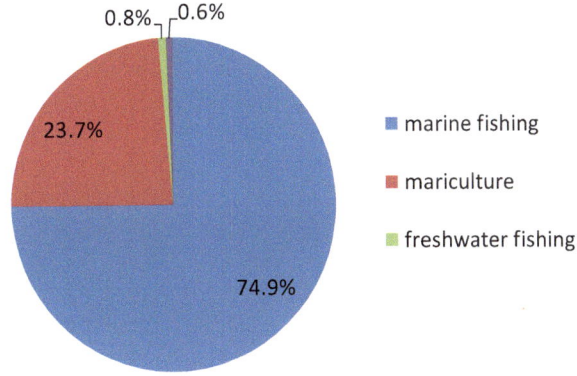

Table 3.2 Marine capture production structure of Japan in 2016

Item	Number
Number of fishery workers (persons)	180,990
Number of fishing vessels	79,563
Production (10,000 tons)	371.5
Per capita production (tons/person)	20.53

Table 3.3 Ratio of capture production to aquaculture production in Japan from 1990 to 2016 (Chen and Zhou 2018)

Year	Capture:Aquaculture
1990	87.60:12.40
2000	79.80:20.20
2005	78.20:21.80
2010	78.30:21.70
2015	76.40:23.60
2016	75.50:24.50

structure, the number of marine capture workers in Japan was approximately 181,000 in 2013, with a total of 79,600 fishing vessels, a fisheries production of 3.715 million tons, and a per capita production of fishermen of 20.5 tons (Table 3.2). In short, the productivity level in coastal and inshore fishing grounds is relatively high.

In terms of marine fisheries, Japan's marine capture production in 2016 was approximately 3.264 million tons, accounting for 4.1% of the global total and ranking sixth in the world. In the total capture fisheries production, the coastal and inshore production was 2.93 million tons, and the deep-sea fisheries was 334,000 tons, accounting for 89.8% and 10.2% of marine capture production, respectively. The marine aquaculture production was 1.033 million tons (approximately 750,000 tons of algae and shellfish, comprising 72.6% of marine aquaculture production), accounting for 3.6% of the global marine aquaculture production, ranking 12th in the world.

In terms of inland fisheries, Japan's production is 63,000 tons (28,000 tons of freshwater capture production and 35,000 tons of freshwater aquaculture production), accounting for 1.4% and 0.1% of Japan's fisheries production and global inland fisheries production, respectively. Although Japan's inland fisheries production accounts for a relatively low proportion, the river fisheries, lake fisheries, and freshwater aquaculture have formed a unique local dietary style, leisure tourism resources, and representative fish species with traditional characteristics, which has played a role in promoting local economic development.

Since the 1990s, Japan's capture production has dropped from 10 million tons to approximately 3 million tons at present. The proportion of aquaculture production has increased, and the ratio of capture to aquaculture production has changed (Table 3.3). In 2016, the ratio of capture to aquaculture production reached 75.50:24.50, but the vast majority originated from marine capture production, accounting for 74.9% of the total fisheries production. This statistic indicates that Japan is a country dominated by marine capture. Its deep-sea fisheries, except that from the high seas, mostly occur at the 200-nautical-mile exclusive economic zones of Pacific island countries and African countries. Inshore fisheries conducted on the fishing grounds are mainly carried out within the framework of intergovernmental agreements with neighboring countries, with production accounting for approximately 59.3% of the marine capture production. Coastal fishery workers account for approximately 80% of the total number of fishery workers, and coastal fisheries production accounts for approximately 30.5% of marine capture production. In the decade from 1990 to 2000, the proportion of capture in total fisheries production decreased by 7.8%, ranking third among China, Indonesia, India, the USA, and Japan, and this figure fell by 4.3% from 2000 to 2016, ranking fourth among the above five countries. The results show that the capture production in Japan is expected to continue this downward trend due to such factors as resources, labor force, and production costs and that the aquaculture industry has recovered with improved production to cater to increasing demand from the international market.

3.3.4 Development Trends of Fisheries in Japan

Japan is one of the most developed countries with regard to fisheries, and the aquatic products provide 40% of the domestic animal protein demand. Although its self-sufficiency rate recovered from 53% in 2000 to 56% in 2016, Japan still faces a situation of decreasing production, decreasing consumption of aquatic products, a declining fisheries production structure, and economic recessions in fishing villages. In this context, it is increasingly arduous to fulfill the target task of achieving a self-sufficiency rate of 60%.

In the future, Japan will adhere to its fisheries policy. Concrete measures to implement this fisheries policy are illustrated as follows: to develop coastal-oriented fisheries and fishing villages; to strengthen the combined use of infrastructure in fishing villages, fishing ports, and fishing grounds; to promote the proper management of aquatic resources and the growth of fisheries industrialization; to improve international resource management and overseas fisheries cooperation mechanisms; to establish an aquaculture system with less environmental load; to maximize sustainable utilization of aquatic resources; to focus on improving the self-sufficiency rate of aquatic products and adjusting the fisheries production structure; to ensure fisheries stability and production efficiency; to cultivate operating entities with international competitiveness; to give full play to the multifunctional role of fishing villages; to establish a dynamic employment structure of the fisheries labor force; to enhance the income of fishery workers; and to achieve a stable supply of aquatic products and the healthy and sustainable development of fisheries.

3.4   Fisheries in the USA

3.4.1   *Natural Environment and Conditions of Fisheries*

3.4.1.1   Natural Environment of Marine Fisheries

The mainland USA, which has coasts on three sides, faces the Atlantic Ocean in the east, the Pacific Ocean in the west, and the Gulf of Mexico on the south, meeting Canada on the north, and is 4500 km wide from east to west and 2700 km long from north to south. The country has a coastline of approximately 22,700 km, a continental shelf area of 2.307 million km^2, and a 200-nautical-mile exclusive economic zone area of 336 km^2.

The USA's waters can be roughly divided into three areas. (1) The eastern coastal area is dominated by a temperate monsoon climate, characterized by four distinct seasons and severely cold winters, with the lowest temperature at −20 °C. The eastern coastal area is muggy in summer, when temperatures of above 40 °C are common and tornadoes and thunderstorms often occur. In the north, New England has a colder mid-temperate oceanic climate with less precipitation. The northcentral Mid-Atlantic region has a temperate subhumid climate, with moderate temperatures and higher precipitation. Virginia and Carolina in the southcentral part have a warm temperate climate with more precipitation. In the south, Florida and Georgia have a subtropical monsoon climate. The western part has a local climate in Appalachian mountainous areas. The bottom sediments are mainly argillaceous and silty. (2) The western coastal area is mostly oceanic climate and Mediterranean climate. Affected by the warm and humid air flow from the ocean, the summer is hot with little precipitation, and the winter is warm and moist. The average temperature in January is above 4 °C, and the average temperature in July is approximately 20–22 °C, with an average annual precipitation of approximately 1500 mm. The western coast of the mainland with a latitude near 30°N in the south is controlled alternately by subtropical high barometric pressure and westerly wind, with more precipitation in winter and less in summer. The western coast of the mainland with a latitude between 40°N and 60°N in the north is controlled by westerly wind and is mild and humid all year round with stable precipitation. The bottom sediments are mainly sandy and are not pelitic. (3) The Gulf of Mexico, located in the tropics and subtropics, is hot and rainy with a large amount of precipitation. The highest temperature comes in August and can be above 28 °C, and the lowest temperature comes in February and is approximately 12 °C in the north and 22 °C in the south. Strong northerly winds blow mainly in winter, and hurricanes often hit in summer, with an average annual precipitation of approximately 1500 mm. Warm water in the Gulf flows out of the Florida Strait, constituting an important source of the Gulf Stream. The surface temperature of the Gulf is relatively high, as high as 29 °C in summer and as low as 20–24 °C in winter, which is 2–3 °C higher than that of the Atlantic Ocean at the same latitude. The seawater salinity is 36.0–36.9, but when the Mississippi River reaches its maximum flow, the salinity of the inshore waters within a range of

30–50 km is as low as 14–20. The warm Caribbean Current flows through the Yucatan Strait into the Gulf of Mexico, forming a clockwise ocean current that flows out of the Florida Strait into the Atlantic Ocean and makes up the preliminary Gulf Stream. The bottom sediments mainly consist of stone and sandy mud.

3.4.1.2 Natural Environment of Inland Fisheries

The inland waters of the USA cover an area of approximately 370,000 km^2. The national annual average surface runoff is approximately 2.9 trillion m^3, accounting for approximately 6.3% of the global total. This factor is influenced by topography and climate, water network density, water system size, water supply, and water volume, and its seasonal changes are distributed very unevenly.

The Rocky Mountains constitute the main watershed of the national water network, followed by the Appalachian highlands in the east and the low moraine ridges in the north. The rivers originating from these three sources alone flow in different directions or converge into water systems with different sizes and flow into the ocean. There are three major water systems in the USA: the rivers to the east of the Rocky Mountains that flow into the Atlantic Ocean are the Atlantic water systems, mainly including the Mississippi River, the Connecticut River, and the Hudson River; rivers that flow into the Pacific Ocean are the Pacific water systems, mainly including the Colorado River, the Columbia River, and so on; and the great lakes group in central and eastern North America – the Great Lakes – are the third-largest water system and the largest freshwater body in the world.

3.4.2 Fish Species

3.4.2.1 Marine Fishery Resources

The USA, located in the northeast Pacific Ocean and the northwest Atlantic Ocean, is rich in fishery resources and has good fishing grounds in the coastal areas of the east, west, and south. Its living marine resources account for 15% of the world total, with 275 families and 450 fish species. Additionally, there are approximately 120 fish species recorded in the fishery statistics and market circulation sales.

As for the status of fishery resources, there are a large number of fish species in the northeast and southeast seas on the Atlantic side due to either an excessive fishing intensity or insufficient resources. In the coastal areas of New England and Georgia, the fishery species mainly include herring, sardine, scallop, clam, cod, pollock, flounder, and so on. In the shallow waters of the mid-Atlantic region, the fish species are mainly scallop, pogy, flounder, and so on. In Chesapeake Bay, the fish species mainly include oyster, clam, blue crab, fish, and so on. In the South Atlantic region, the fish species mainly include mullet, oyster, shrimp, blue crab, mackerel, spiny lobster, saury, and so on. In the southwest and northwest seas and the Alaskan waters along the Pacific side, fish species are not diversified, and the species with high economic value are mainly pollock, salmon, tuna, flounder, and crab. In Alaskan waters, the fish species are mainly pollock, salmon, Dungeness crab, flounder, herring, scallop, and so on. In Washington and Oregon waters, the fish species are mainly albacore tuna, Pacific whiting, Dungeness crab, sea bass, salmon, oysters, and so on. In the waters of northern California, the fish species mainly include salmon, flounder, Dungeness crab, herring, and so on. In the waters of central California, the fish species mainly include shrimp, abalone, sea bass, albacore tuna, Dungeness crab, and so on. In the waters of southern California, the fish species mainly include bluefin tuna, yellowtail tuna, shrimp, lobster, dorado, pike, white jewfish, mackerel, anchovy, and so on.

According to statistics on the exploitation and utilization of marine fishery resources, Alaska pollock is the only fish species with an annual catch of more than 1 million tons; herring is the only fish species with a catch of 700,000–800,000 tons; only 2 species, the Pacific whiting and drumfish, have catches of 200,000–300,000 tons; only 4 species have catches of 100,000–200,000 tons, which are sockeye salmon, yellowfin sole, crab, and shrimp, and approximately 14 species have catches of 10,000–100,000 tons, mainly including grouper, Atlantic herring, pink salmon, mackerel, arrowtooth flounder, salmon, squid, cuttlefish, bastard halibut, tuna, ray, sablefish, shellfish, mollusk, and so on.

3.4.2.2 Inland Fishery Resources

The inland waters of the USA are abundant in fishery resources, with approximately 900 species of freshwater fish. In the 1970s, the rivers and soil in the USA were polluted, and some rivers and fishponds in the southern states were covered with algae and aquatic plants. To improve water quality, the US government introduced cyprinid fish from Asia such as grass carp, carp, and silver carp. However, their strong reproductive ability and survivability has led to a sharp decline in the number of native freshwater fish in the USA; thus, the Asian carps have been classified as the "most dangerous exotic fish species" in the USA.

In terms of fish species, tilapia, shuttle fish, shark catfish (*Pangasianodon gigas*), blue catfish, largemouth bass, bluegill, and hornyhead chub (*Nocomis biguttatus*) are common species. Shrimp and shellfish resources, such as *Macrobrachium rosenbergii*, Florida crayfish, red arowana, red swamp crayfish (*Procambarus clarkii*), zebra mussel (*Dreissena polymorpha*), and pink scallop are also abundant in the USA. Americans rarely eat freshwater fish, and the government stipulates that fish less than 50 g and more than 1500 g from recreational fishing shall be released, which makes the fish resources in freshwater areas very abundant.

3.4.3   Current Status of Fisheries Development

The USA is one of the major fishery countries in the world, and the 1980s witnessed rapid fisheries development in the USA. The total fisheries production in 1987 exceeded 6 million tons, an increase of 57% compared with 1980, and the fisheries production in 2016 was approximately 5.375 million tons, ranking fifth in the world, accounting for 2.7% of the world total.

As for the fisheries production structure, marine fisheries are the priority in the USA, with its marine capture production ranking second to Indonesia, accounting for 91.3% of the total (Fig. 3.4), which is 44.9% higher than the global marine capture. However, its aquaculture production only accounts for 8.3% of the total, 66.2% and 38.5% lower than that of China and the world, respectively. These statistics indicate that marine capture has a highly significant status in the USA. In spite of some improvements in aquaculture, there is limited potential for aquaculture development due to the influence of environmental protection and other factors. As for the capture production structure, in 2016, there were approximately 118,000 marine fishing workers in the USA, with 10,000 fishing vessels, 4.897 million tons of fisheries production, and 41.5 tons of per capita production of fishermen (Table 3.4). The productivity was approximately two times higher than that of Japan, which was attributable to the contribution from commercial fishing in the USA.

As for the marine fisheries, the production of marine capture and aquaculture in the USA accounted for 94.7% of the total in 2016, of which marine capture production was approximately 4.897 million tons, accounting for 6.2% of global production, a decrease of 6.1% compared with 2000. Alaska plays an important role

Fig. 3.4 Fisheries production structure of the USA in 2016 (FAO 2018)

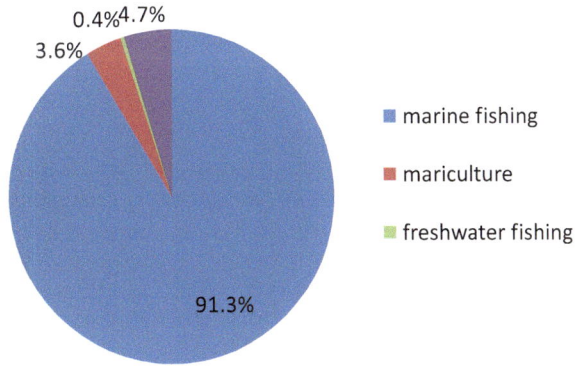

0.4% 4.7%
3.6%

91.3%

■ marine fishing

■ mariculture

■ freshwater fishing

Table 3.4 Marine capture production structure of the USA in 2016

Item	Number
Number of fishery workers (persons)	118,000
Number of fishing vessels	N/A
Production (10,000 tons)	489.7
Per capita production (tons/person)	41.5

Table 3.5 Ratio of capture production to aquaculture production in the USA from 1990 to 2016 (Chen and Zhou 2018)

Year	Fishing yield:Aquaculture production
1990	94.60:5.40
2000	91.20:8.80
2005	90.60:9.40
2010	90.00:10.00
2015	92.30:7.70
2016	91.50:8.50

in US fisheries, with approximately 2.58 million tons of marine capture production in 2016, accounting for 58% of the national total, ranking first for 20 consecutive years. Marine aquaculture production reached 194,000 tons (including 16,000 tons of salmon and trout, 30,000 tons of clam, 141,000 tons of oyster, comprising 72.8% of the marine aquaculture production), accounting for approximately 3.6% of the total fisheries production.

As for the inland fisheries, the production in 2016 was approximately 275,000 tons, accounting for 5.1% of the total, of which freshwater aquaculture production was 253,000 tons, accounting for 4.7% of the total. Salmon production was 112,000 tons, accounting for 44.3% of the freshwater aquaculture production. The freshwater capture production was 22,000 tons, accounting for approximately 0.4% of the total.

In 2016, the ratio of capture to aquaculture production in the USA reached 91.50:8.50 (Table 3.5), of which marine capture occupied an overwhelming position, maintaining above 90% for a long time, and the aquaculture potential remains to be developed. Some reasons contribute to the primary position of marine capture in the USA: located between the Atlantic Ocean and the Pacific Ocean, the USA has an abundance of marine fishery resources; the utilization of aquaculture resources is deficient due to many factors, and the production is declining year by year. The utilization of angling in recreational fisheries has been well developed and strictly managed, which is conducive to the conservation of fishery resources. In the decade from 1990 to 2000, the proportion of total capture production decreased by 3.4%, ranking fourth among China, Indonesia, India, the USA, and Japan. From 2000 to 2016, the proportion of capture production increased by 0.3%, ranking fifth among the above five countries. This statistic indicates that with the enhancement of fishery resources management, the production of marine fishery resources in the USA has gradually increased, and the production of high-value fish consumed in the international market has also increased.

3.4.4 Development Trends of Fisheries in the USA

The USA is blessed with extensive sea areas, and the living marine resources in the northwest sea area and Alaska sea area are in relatively good condition, especially in Alaska, where the stable capture production accounts for a large proportion of the national fisheries production. Moreover, to prevent overfishing and achieve sustainable development of fisheries, the US government has been enhancing fisheries resource conservation and fisheries management to promote the stability of marine capture production. The exploitation level of inland fishery resources is relatively low, but developing recreational fisheries, the priority for fisheries development in the USA, is conducive to protecting ecological diversity.

Some highlights of the fisheries development trend are illustrated as follows. The USA will continue to maintain a good resource status for marine capture and strengthen fisheries management. The commercial fisheries will maintain the principal position in fisheries production. Freshwater aquaculture will continue to maintain the primary position. As for marine aquaculture, environmental protection is still the major concern, and priority is to be given to shellfish farming without feeding and salmon farming. There will be no significant increase in marine capture production, but the recreational fisheries industry will remain in a leading global role.

3.5   Fisheries in Indonesia

3.5.1   Natural Environment and Conditions of Fisheries

3.5.1.1   Natural Environment of Marine Fisheries

Indonesia is the largest archipelagic country in the world with extensive sea areas, numerous islands, and widely distributed reefs. Indonesia has a coastline of approximately 810,000 km, 13,667 islands, large and small, between the Pacific Ocean and the Indian Ocean, and covers an extensive sea area of 5.8 million km^2, where the exclusive economic zone is 3.1 million km^2, ranking third only to Australia and the USA.

Indonesian waters can be divided into three sea areas. (1) The Java Sea, located among Java Island, Sulawesi Island, Kalimantan Island, and Sumatra Island in the western Pacific area, borders Borneo (Kalimantan) to the north; connects the south end of Makassar Strait to the northeast; is adjacent to Celebes, Flores, and the Bali Sea to the east; connects Java to the south; borders Sunda Strait to the southwest; is bounded by Sumatra to the west; and borders Bangka Island and Belitung Island to the northwest. This sea is approximately 1450 km wide from east to west and 420 km wide from north to south, covering an area of 433,000 km^2, with an average water depth of 50 m. The sediment particles around the coast are relatively fine, the fine sand is widely distributed, and the particles in the central basin are coarse. (2) The Arafura Sea, an interisland sea on the eastern edge of the Indian Ocean, is located between New Guinea (Irian) and the northern coast of Australia. The Arafura Sea is approximately 1280 km wide from east to west and 560 km wide from north to south, covering an area of 1.032 million km^2, with an average water depth of 197 m and a maximum depth of 3680 m. The entire sea area lies on the continental shelf.

There are many coral reefs along the southern coast, and four reefs and shoals are distributed in an east-west direction along the northern boundary. (3) The Banda Sea, located in the western part of the South Pacific Ocean, is surrounded by the southern islands of Moluccas. The Banda Sea is connected with the Flores Sea to the west, the Savu Sea to the southwest, the Timor Sea to the south, the Arafura Sea to the southeast, and the Selan Sea and Molucca Sea to the north, covering an area of 470,000 km^2, with an average water depth of 3064 m. The sea area lies on an unstable active zone of the crust, with many volcanic islands.

Crossing through the equator, Indonesia is hot and rainy all year round, with abundant rainfall and balanced temperature, and the annual average temperature is 25–27 °C, with an average annual precipitation of more than 2000 mm. As the surface current of this sea area corresponds to the wind current, the annual surface water temperature is basically maintained at 24–29 °C.

3.5.1.2 Natural Environment of Inland Fisheries

Indonesia has a large number of rivers, covering the islands of Kalimantan, Sumatra, Papua, and Celebes, as well as Java, Bali, and Nusa Tenggara. Indonesia is blessed with abundant waters and has an area of internal waters of approximately 13.85 million ha, including 12 million ha of rivers and floodplains, 1.8 million ha of lakes, and 5 million ha of artificial lakes and reservoirs. Affected by the monsoon in the northern hemisphere, the northern part has abundant precipitation from July to September each year, while the southern part, affected by the southern hemisphere monsoon, has abundant precipitation from December to February of the following year. The main rivers are the Solo River, the Barito River, the Kapuas River, and the Mahakam River, and the main lakes are Lake Toba, Lake Maninjau, Lake Cara, Tempe Lake, Lake Towuti, Panai Lake, and so on.

The inland waters of Indonesia can be divided into the following areas. (1) The Solo River, the longest river in Java, is 560 km long. The Solo River originates from the slopes of the Lavu volcano and the southern part of the Sewu Mountains and flows northward and turns eastward into the sea in northwestern Surabaya. (2) The Barito River, the southern river of Kalimantan, has a total length of 890 km, with an average annual runoff of 5500 m^3/s. The Barito River originates from the northern border mountains and runs into the Java Sea. This river has a basin area of 100,000 km^2, and two-thirds of the flow is navigable. (3) The Kapuas River, the main waterway in the western part of Borneo, has a tropical rainforest climate, with an abundant annual rainfall of over 2600 mm and an annual average temperature of 25–27 °C. The Kapuas River is 11,431 km long and has a basin area of 98,000 km^2. The river networks in this basin are densely distributed with numerous rivers, lakes, and swamps. This river is located at the equator and 109–114°E. The topography is high in the east but low in the west, the terrain features plains and swamps, and there are mountains and hills at the upstream headwaters. (4) The Mahakam River, an important river in East Kalimantan, has a total length of approximately 650 km. The Mahakam River originates from the Ilan Mountains at the border, with many long tributaries, and flows into the inland basins of East Kalimantan Island, forming many lakes and large swamps. The southeastern flow is divided into four cross flows that flow into Makassar Strait. (5) Lake Toba, located in the Mata Plateau in northern Sumatra, is the largest freshwater lake in Indonesia and also the most well-known plateau lake in the world, with high temperature and rainfall all year round, although with no great temperature variation. The rainfall significantly varies: the annual precipitation on the west coast is 3000 mm, while as high as 4500–6000 mm in the mountainous areas, 2000–3000 mm from the eastern slope of the mountains to the coastal plain, and 1500–1700 mm at the northern and southern ends of the island.

3.5.2 Fisheries Species

3.5.2.1 Marine Fishery Resources

There are abundant fishery resources around the islands of Indonesia, and the living resources fall under the category of tropical flora. There are more than 230 species available for fishing, including 65 species of great economic value. There are more than 80 shrimp and crab species, 100 lamellibranch species, 1500 gastropod species, 65 sea cucumber species, and 555 algal species.

According to the water layers that fish inhabit, scope of the sea areas where fish are distributed and characteristics of relevant fishery resources, fishery resources can be classified as follows:

1. The pelagic species, which mainly include tuna, skipjack, yellowfin tuna, Spanish mackerel, chub mackerel, herring, sardine, amberfish, round herring, squid, flying fish, and so on.
2. The demersal species, which mainly include *Serranidae*, *Lutjanidae*, *Trichiuridae*, *Leiognathidae*, *Nemipteridae*, *Priacanthidae*, *Stromateidae*, *Siluridae*, *Ariidae*, *Drepanidae*, *Mullidae*, *Psettodidae*, *Cynoglossidae*, and so on.
3. Highly migratory fish species, which are mainly bluefin tuna, yellowfin tuna, bigeye tuna, albacore tuna, southern bluefin tuna (*Thunnus maccoyii*), skipjack, plain bonito, bullet tuna (*Auxis rochei*), northern blue tuna (*Thunnus tonggol*), and so on.
4. The reef fish, which mainly include *Pomacentridae*, *Caesionidae*, *Scaridae*, *Holocentridae*, *Acanthuridae*, *Siganidae*, *Serranidae*, *Lethrinidae*, *Labridae*, *Lutjanidae*, *Priacanthidae*, and so on.
5. Shrimp and crabs, which mainly include approximately 40 species of *Penaeidae*, as well as the Indian prawn (*Penaeus indicus*), *Penaeus monodon*, *Metapenaeus ensis*, *Parapenaeopsis hardwickii*, lobster, and mud crab (*Scylla serrata*).
6. Mollusks, which mainly include clam, mud clam, Amusium, stromb, oyster, giant clam, top shell (*Trochid*), pearl shell, sea cucumber (*Holothurians*), abalone, calamari, cuttlefish, and so on.

7. Algae, which are mainly red algae, brown algae, and green algae, but only 55 species are edible or for medical use.

According to the catch data from FAO, Indonesia has abundant tuna resources and has the largest tuna-like production fishing ground with an annual catch of more than 1 million tons. Shrimp and short mackerel have catches of 200,000–300,000 tons. Additionally, 7 species have catches of 100,000–200,000 tons, including squid, long-jaw anchovy, golden sardine, Spanish mackerel, crab, bream, and horse mackerel. In addition, 8 species have catches of 50,000–100,000 tons, including nearshore bream, yellow-striped mackerel, drumfish, goatfish, Indian mackerel, mullet, lithosporic, and black pomfret. Additionally, 4 species have catches of 10,000–500,000 tons, including the Bali sardine, pomfret, surf clam, and red algae. Therefore, there is still great potential for the exploitation of fishery resources in Indonesia.

3.5.2.2 Inland Fishery Resources

Inland open waters in Indonesia are home to diverse fish species, with a total of more than 800 species. There are 368 fish species in Sumatra and Kalimantan, 798 fish species in the Sunda Continental Plain, 68 species in the Wallacea Islands, and 106 species in the Sahur Continental Plain, over 200 species in the Kapuas River, and more than 104 species in the Barito River. There are catfish, milkfish, carp, freshwater shrimp, and endemic fish around some islands, but the fishing potential has not yet been realized.

Aquaculture methods in Indonesia include net cages, floating net cages, and fish cultivation in paddy fields. Production of brackish water aquaculture is the highest, and these farms account for approximately 60% of the inland aquaculture water surface, followed by freshwater aquaculture. Freshwater aquaculture species include common carp, Nile tilapia, Nile carp (*Osteochilus hasseltii*), snakehead carp, silver carp, bighead carp, grass carp, and so on, among which common carp is the most popular, mainly farmed in the West Java and North Java regions, and its development and utilization has been rapidly promoted for food security, employment opportunities, and foreign exchange earnings.

3.5.3 Current Status of Fisheries Development

Indonesia is the second largest fisheries producer in the world. In terms of capture, the vast exclusive economic zones and deep water depth make migratory fish resources such as salmon and tuna abundant, which provides favorable conditions for marine capture.

According to fishery statistics released by the FAO, fisheries-oriented provinces in Indonesia, and other institutions, it was estimated that the total fisheries production of Indonesia in 2016 was approximately 11.542 million tons (excluding approximately 11 million tons of algal production), accounting for 11.5% of the world total, an increase of 88.6% compared with 2004. In terms of capture, Indonesia has the second largest capture production after China, accounting for 7.2% of the global total capture production. The total aquaculture production exceeds 5 million tons, ranking third in the world.

As for the fisheries production structure of Indonesia, capture occupies an important position in fisheries production, accounting for 56.6% of the total fisheries production (Fig. 3.5), an increase of 40.6% compared with 2004. The proportion of capture production in fisheries production is higher than that of aquaculture, with a difference of 13.4%. The proportion of marine capture production is 9.7% higher than that of aquaculture. This statistic shows that marine fisheries are the major contributor to fisheries production in Indonesia, but aquaculture is developing relatively rapidly. As for the capture production structure, in 2015, there were approximately 460,000 marine fishing workers in Indonesia, 568,000 fishing vessels, 6.109 million tons of capture production, and 13.3 tons of per capita production of fishermen (Table 3.6), with a medium level of productivity.

In terms of marine fisheries in Indonesia, the marine fisheries production in 2016 was 7.63 million tons, accounting for 7.1% of the global total, which was several times higher than that in 2000. Specifically, the marine capture production was approximately 6.11 million tons, accounting for 7.7% of the global total. The production of tuna, skipjack, and swordfish was 1.318 million tons, accounting for 21.6% of marine capture production; other fish production was 1.322 million tons, accounting for 18.5%; the shrimp and crab production was 429,000 tons, accounting for 7.0% of marine capture production; the mollusk production was 264,000 tons, accounting for 4.3% of marine capture production. Marine aquaculture has

Fig. 3.5 Fisheries production structure of Indonesia in 2016 (FAO 2018)

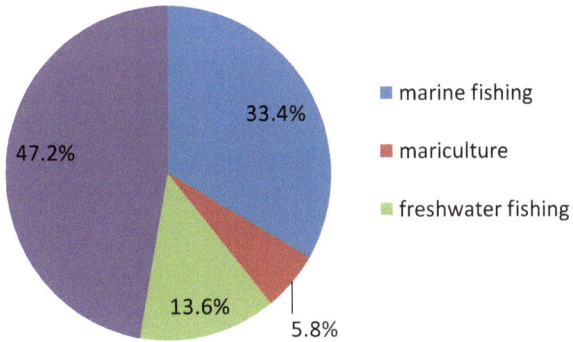

Table 3.6 Marine capture production structure of Indonesia in 2016

Item	Number
Number of fishery workers (persons)	460,000
Number of fishing vessels	568,329
Production (10,000 tons)	610.9
Per capita production (tons/person)	13.3

Table 3.7 Ratio of capture production to aquaculture production in Indonesia from 1990 to 2016 (Chen and Zhou 2018)

Year	Capture:Aquaculture
1990	81.50:18.50
2000	80.80:19.20
2005	70.50:29.50
2010	46.20:53.80
2015	30.10:69.90
2016	56.70:43.30

developed rapidly, with a production of approximately 1.52 million tons, mainly including fish, shrimp, shellfish, and algae, accounting for 19.9% of marine fisheries production. In 2016, the fish production was 864,000 tons, and the shrimp production was 644,000 tons.

As for the inland fisheries in Indonesia, the production in 2016 was approximately 3.91 million tons, accounting for 33.9% of the total fisheries production, an increase of 3.5 times compared with 2000, making it the country with the third largest inland fisheries production after India. Specifically, freshwater aquaculture production was approximately 3.48 million tons, accounting for 6.8% of the global freshwater aquaculture production; inland capture production was 432,000 tons, accounting for 3.7% of the global total, ranking sixth in the world, and showing a steady growth trend in recent years.

In 2016, the ratio of capture to aquaculture production was 56.70:43.30 (Table 3.7). Compared with 2000 and 2010, the capture production increased by 58.2% and 22.1%, respectively, but the proportion of capture decreased by 24.1% and increased by 10.5%, respectively. The aquaculture production increased by 16.8% and 2.6 times, respectively, and the proportion of aquaculture increased rapidly. Freshwater aquaculture production was more than 128.9% that of marine aquaculture. In terms of aquaculture species, fish and shrimp aquaculture developed rapidly, within which fish culture accounted for 69.6% of the total and shrimp culture accounted for 13.8% of the total. In the decade from 1990 to 2000, the proportion of capture in total fisheries production decreased by 0.7%, ranking fifth among China, Indonesia, India, the USA, and Japan. From 2000 to 2016, this figure fell by 24.1%, while still ranking first among the above five countries. This statistic shows that while the capture capacity has increased since the beginning of the twenty-first century, the aquaculture industry has maintained a good development momentum.

3.5.4 Development Trends of Fisheries in Indonesia

In Indonesia, fisheries are still conducted in the traditional way. Aquaculture, especially finfish freshwater aquaculture, has developed rapidly, which has played a positive role in providing employment opportunities and income enhancement for residents along inland waters. As a major marine capture country, Indonesia is also encountering fisheries challenges, with illegal, unregulated, and unreported fishing (IUU fishing) frequently occurring in its waters, which brings about severe challenges in the export, sale, and quality of fisheries products. At the same time, due to the characteristics as an archipelago country, its fisheries development is not balanced in most regions.

In the future, Indonesia will prioritize the improvement of fisheries human resources, technological capability, and infrastructure. The top priority will go to the development of inland aquaculture, which is believed to ensure that rural food needs are met and that low-priced animal protein is available for the poor. Some concrete measures are to be taken by the Indonesian government to enhance fisheries development, such as developing high-priced, export-oriented aquaculture species, formulating standardized and regulated industrial development policies to reduce poverty and increase the income of fishermen, and promoting sustainable fisheries development.

3.6 Fisheries in India

3.6.1 Natural Environment and Conditions of Fisheries

3.6.1.1 Natural Environment of Marine Fisheries

India has a coastline of approximately 8118 km, most of which belongs to the peninsula in southern Asia, extending into the Indian Ocean, bordering the Arabian Sea in the southwest and the Bay of Bengal in the east and southeast. The Lakshadweep Islands are coral atoll islands on the southwestern coast, and the world-famous Andaman and Nicobar Islands are located in the volcanic chain islands in the Andaman Sea. India has a continental shelf of 530,000 km^2 and an exclusive economic zone of 2.02 million km^2.

In India, sea areas are mainly categorized into the following areas. (1) The Bay of Bengal, located in the northern part of the Indian Ocean, borders the Indian Peninsula to the west, the Indochina Peninsula to the east, and Myanmar and Bangladesh to the north, connects Sri Lanka and Sumatra to the south by connecting the Indian Ocean, and borders the Gulf of Siam and the South China Sea through the Strait of Malacca. The Bay of Bengal is the largest bay in the world, with a width of approximately 1600 km, an area of 2.17 million km, and a water depth of 2000–4000 m, although the southern part has the deeper water, with salinity ranging from 30‰ to 34‰. Many major rivers in India flow from west to east before flowing into the Bay of Bengal, including the Padma River, the Meghna River, and the Jamuna River. There are numerous islands in the Bay, including the Andaman Islands, the Nicobar Islands, and the Melgi Islands. (2) The Andaman Sea, located in the southeastern waters of the Bay of Bengal, faces the northeastern Indian Ocean, with a width of approximately 1100 km from north to south, a width of 600 km from east to west, and an area of approximately 798,000 km^2. To the west of the Andaman

Sea are the Andaman Islands of India, to the east is the Malay Peninsula, to the north are Thailand and the Irrawaddy Delta of Myanmar, and to the southeast is the Strait of Malacca. This sea is an important waterway between the South China Sea and the Indian Ocean. The depth of the Andaman Sea varies greatly, but approximately 5% of the sea areas are deeper than 3048 m, and the deepest areas are in a series of deep-sea valleys near the Andaman Islands, with a water depth of more than 4420 m. The surface water temperature fluctuates slightly, from 27.5 °C in winter to 30 °C in summer. (3) The Arabian Sea, located between the Arabian Peninsula in southern Asia and the Indian Peninsula, is part of the Indian Ocean. Its northern part is the Persian Gulf and the Gulf of Oman, and its western part joins the Red Sea by passing through the Gulf of Aden. This sea covers an area of 3.86 million km^2, with a maximum width of approximately 2400 km and a maximum depth of 4652 m. The Arabian Sea is located in the tropical monsoon region with high temperatures. The surface water temperature in the central sea area is often above 28 °C in June and November, and the temperature in January and February decreases to 24–25 °C. The water temperature near the Arabian Peninsula is as high as 30 °C due to the influence of dry and hot air flow on land. From November to March of the following year, the northeast monsoon blows with little precipitation, and the southwest monsoon blows with abundant precipitation from April to October. Tropical cyclones often occur at the turn of summer and autumn, accompanied by huge waves and heavy rainstorms. The sea current varies with the monsoon and is clockwise in summer under the influence of the southwest monsoon and counterclockwise in winter under the influence of the northeast monsoon. The seawater salinity is generally less than 35‰ in the rainy season and more than 36‰ in the dry season. (4) The Laccadive Sea (the Lakshadweep Islands), located between the Maldives and Sri Lanka and India, is to the southwest of Karnataka, the west of Kerala, and the south of Tamil Nadu. The water temperature is quite stable throughout the year, with an average of 26–28 °C in summer and 25 °C in winter. The salinity of the central and northern parts is 34‰, and that of the southern part is 35.5‰. The coast is sandy, but the deeper part of the sea is covered with silt. There are many coral reefs in the sea, of which the Lakshadweep Islands are composed of atolls with 105 coral species.

3.6.1.2   Inland Fishery Resources

India is rich in inland water resources, with rivers and canals totaling 195,000 km in length. There are 14 large rivers, 44 medium-sized rivers, and countless small rivers in India. The estuaries cover an area of 300,000 ha, the saltwater and lagoon area is 190,000 ha, the area of highland lakes is 720,000 ha, and the area of reservoirs is 3.15 million ha, including 561 large reservoirs, 180 medium-sized reservoirs, and 19,134 small reservoirs, with areas of 1.14 million ha, 527,000 ha, and 1.485 million ha, respectively. In recent years, with the successive construction of various dams in rivers, streams, and other waterways, the number of reservoirs has been increasing year by year. The floodplain wetlands cover an area of 200,000 ha and are mainly distributed in Assam, West Bengal, and Bihar. The abundant nutrients and biological bait make these water bodies very favorable for the development of aquaculture-based fisheries.

There are many rivers in India, including the Indus River, the Ganges River, the Brahmaputra River, and five major water systems along the eastern, western, southern, and northern coasts. The first three water systems originate in the Himalayas, supplied by snowmelt and glaciers. The interannual and intra-annual changes in water volume are relatively stable, with the total runoff accounting for approximately 63% of the country's total runoff. The latter two water systems are affected by the monsoon, with large flows in the rainy season and small flows in the dry season, mainly dependent on groundwater.

The main rivers in the country are as follows. (1) The Ganges River, whose basin area in India, is 861,000 km^2, accounting for 79.3% of the total area of the Ganges River Basin. The Ganges River is integrated with the YarlungZangbo River through a complex common diversion system, flowing through Uttar Pradesh, Madhya Pradesh, Bihar, Rajasthan, Himachal Pradesh, Delhi, and Arunachal Pradesh and finally flowing into the Bay of Bengal. The hydrological cycle of the Ganges River Basin is affected by the southwest monsoon, resulting in obvious seasonal flows, and the rainfall from June to September accounts for approximately 84% each year. (2) The Indus River, one of the longest rivers in Asia, originates from the Qinghai-Tibet Plateau near Lake Manasarovar, flows through the Ladakh region of Kashmir, and flows southward along the entire territory of Pakistan into the Arabian sea, covering an area of 1.165 million km^2. (3) The Brahmaputra River Basin, the main river in Central and South Asia, is approximately 2900 km long from the origin of the Himalayas in Tibet, China, to its terminal at the Bay of Bengal where it mixes with the Ganges River. The Brahmaputra River has a total length of approximately 1000 km in India, including 725 km in Assam. The Dihang and Brahmaputra River basins cover an area of 292,000 km^2, with an annual average runoff of approximately 200 billion m^3 and an annual sediment discharge of approximately 500 million tons. Affected by the southwest monsoon (from June to September) and cyclones in the Bay of Bengal, this basin has the most rainfall in the world. (4) The Godavari River Basin is the second largest river in India after the Ganges River with a total length of 1465 km. The Godavari River flows eastward through Maharashtra, Teran Ghana, Andhra Pradesh, Chhattisgarh, Madhya Pradesh, and Audisha among others and finally flows into the Bay of Bengal through a vast tributary network. This basin covers an area of approximately 313,000 km^2 and is one of the largest river basins in the Indian subcontinent. (5) The Krishna River Basin, also known as the Krishnaveni River, is the main source of irrigation in Maharashtra, Karnataka, Teran Ghana, and Andhra Pradesh and is the fourth largest river in India. The Krishna River has a total length of approximately 1300 km and covers an area of 259,000 km^2, accounting for 8% of the whole national area. The annual average surface water potential is 78.1 km^2, of which 58 km^2 are available as useable water sources. The area favorable to aquaculture in the basin is approximately 203,000 km^2, accounting for 10.4% of that in India. (6) The Kaveri River Basin, which runs through Karnataka and Tamil Nadu, is the third largest river after

Godavari and Krishna in southern India, covering an area of approximately 81,000 km².

The river ecosystems and fishery resources in the abovementioned basins are seriously threatened by sediment deposition, water extraction, dam construction, and water pollution, and thus, it is necessary to tap the fisheries production potential of these water bodies through developing aquaculture-based fisheries.

3.6.2 Fish Species

3.6.2.1 Marine Fishery Resources

India boasts abundant marine fishery resources, and its warm and fertile inshore waters are among the most productive fishing grounds in the world for prawn, tuna, sardine, mackerel, drumfish, flounder, and various other marine fish. There are approximately 1400 finfish species, 263 species of which are commercially exploitable; 36 prawn species and 34 cephalopod species, of which 15 prawn species and 8 cephalopod species are commercially exploitable; and 356 ornamental fish species.

According to the water depth that fish inhabit, scope of sea areas where fish are distributed, and characteristics of relevant fishery resources, fishery resources can be classified as follows:

(1) The pelagic species, which mainly include chub mackerel, trevally, sardine, oil sardine, anchovy, mackerel, and so on.

(2) The demersal or near-demersal species, which mainly include drumfish, *Regalecus russelii*, squid, cuttlefish, and prawn.

(3) The estuarine fishery resources, which mainly include mullet, perch, *Parequula melbournensis*, and so on.

(4) The highly migratory species, which include herring, tuna, and so on.

Statistics from the Indian Marine Fisheries Agency show that the total annual renewable resources of aquatic products in the waters of the exclusive economic zone are approximately 3.9 million tons. The sardine is the only species with an actual annual catch of nearly 1 million tons; tuna and trevally are the two species with annual catches of 400,000–500,000 tons; 9 species have catches of 100,000–300,000 tons, which include perch, oarfish, skipjack, chub mackerel, oil sardine, prawn, shark, drumfish, and abalone; species with catches of 10,000–100,000 tons are *Parequula melbournensis*, cephalopod, Indian squid, carp, sea crayfish, prawn, and other trash fish.

Under the influence of the monsoon and coastal topography, the western coast in India is a sea area very rich in nutrients. In summer, the pH value of the waters in this sea area is 8–8.3, which is a favorable range for pelagic fish habitat.

3.6.2.2 Inland Fishery Resources

India is blessed with abundant and various inland fishery resources. Cyprinidae fish are the most important species in freshwater aquaculture, mainly including Indian catla, *Labeo udaipurensis*, and mrigal carp (*Cirrhinus mrigala*), accounting for more than 90% of the total aquaculture production.

In the 1970s, the Cyprinidae fish is introduced from China. Silver carp, grass carp, and common carp became the second most important aquaculture species in India. In catfish aquaculture, the Filipino catfish has received extensive attention. The spiny catfish has been farmed in swamps and wastewater in the eastern area. Fish such as giant catfish, low-eyed giant catfish, and tilapia have also been developed in aquaculture in India. The species used for shrimp culture mainly include both marine and freshwater shrimp species, which include the giant river prawn (*Macrobrachium rosenbergii*), monsoon river prawn (*M. malcolmsonii*), giant tiger prawn (*Penaeus monodon*), and white-leg shrimp (*P. vannamei*), and farmed shrimp constitutes the major products for export.

3.6.3 Current Status of Fisheries Development

India is the third largest fisheries producer in the world, occupying an important position in global fisheries. In 2016, the total fisheries production was 10.765 million tons, accounting for 6.3% of the global total. After China, India is the second largest aquaculture country, where aquaculture production accounts for 7.1% of the global total, and capture production accounts for 47% of India's total and 5.6% of the global total, ranking sixth in the world.

As for the fisheries production structure in India, freshwater aquaculture in inland fisheries has played an important role in fisheries production, accounting for 47.2% of the total fisheries production (Fig. 3.6), an increase of 2.2 times compared with 2004; marine capture accounted for 33.4% of the total fisheries production. The proportion of inland fisheries production in the fisheries production structure is much higher than that of marine fisheries, with a difference of 21.4%; the proportion of freshwater aquaculture is more than 7.9% compared with that of marine fisheries, which shows that India's fisheries production structure is mainly focused on inland fisheries at present, and it is a country dominated by freshwater aquaculture. In terms of the capture production structure, in 2015, there were approximately 933,000 marine fishing workers in India, 199,000 fishing vessels, 4.227 million tons of fisheries production, and 4.3 tons of per capita production of fishermen (Table 3.8), which indicates a relatively low level of productivity.

In terms of marine fisheries in India, the production in 2016 was 4.227 million tons, accounting for 3.9% of the global fisheries production, an increase of 1.5 times compared with 2000, within which the marine capture production was approximately 3.6 million tons, accounting for 4.5% of the global production. The herring,

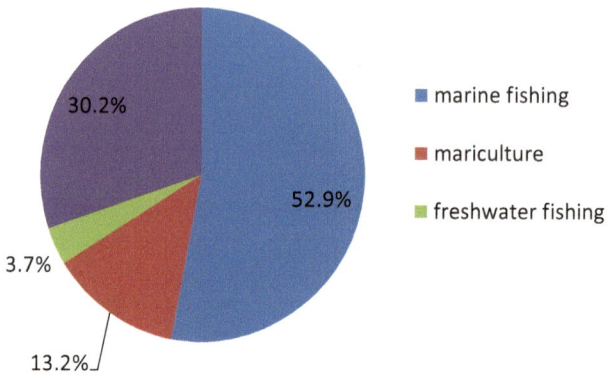

Fig. 3.6 Fisheries production structure of India in 2016 (FAO 2018)

Table 3.8 Capture production structure of India in 2016

Item	Number
Number of fishery workers (persons)	933,124
Number of fishing vessels	199,141
Production (10,000 tons)	422.7
Per capita production (tons/person)	4.3

sardine, anchovy, trash fish, and tuna production was 2.743 million tons, accounting for 76.2% of marine capture production; the shrimp and crab production was 443,000 tons, accounting for 12.3% of marine capture production; and the mollusk production was 420,000 tons, accounting for 11.7% of marine capture production. The marine aquaculture production was 627,000 tons, mainly including shrimp, fish, shellfish, and algae, accounting for 14.8% of marine fisheries production. In 2016, the shrimp production was 521,000 tons, accounting for 83.1% of the aquaculture production; the fish production was 90,000 tons, accounting for 14.4% of the aquaculture production; the seaweed production was approximately 3000 tons, ranking fifth in the world, an increase of three times compared with 2005. Mussel and oyster aquaculture has rapidly developed and has the potential for further development.

In terms of inland fisheries in India, the production in 2016 was approximately 6.538 million tons, accounting for 60.7% of the total fisheries production, an increase of 3.8 times compared with 2000, making India the second largest inland fisheries country after China. Specifically, freshwater aquaculture production was 5.076 million tons, accounting for 9.9% of the global freshwater aquaculture production; inland capture production was 1.462 million tons, accounting for 12.6% of the global production, ranking second in the world, and enjoying a good development momentum in recent years.

The ratio of capture production to aquaculture in 2016 was 39.30:60.70 (Table 3.9). Compared with the beginning of the twenty-first century, the proportion of capture has decreased year by year, while that of aquaculture has increased

Table 3.9 Ratio of capture production to aquaculture production in India from 1990 to 2016 (Chen and Zhou 2018)	Year	Capture:Aquaculture
	1990	73.70:26.30
	2000	49.70:50.30
	2005	55.60:44.40
	2010	55.50:44.50
	2015	48.00:52.00
	2016	39.30:60.70

rapidly, which is conducive to the conservation of marine fishery resources to a certain extent, and the production of freshwater aquaculture exceeded 12.6% of total fisheries production. As for the freshwater aquaculture species, Cyprinidae fish, shrimp, and shellfish have developed rapidly. Among them, common carp production was the highest, followed by barbel and other Cyprinidae fish, which reached 5.066 million tons in 2016, accounting for 88.9% of freshwater aquaculture production. Shrimp production was 521,000 tons, accounting for 9.1% of freshwater aquaculture production. In the decade from 1990 to 2000, the proportion of capture production in total fisheries production decreased by 24%, ranking first among China, Indonesia, India, the USA, and Japan. From 2000 to 2016, this figure dropped by 10.4 percentage points, ranking second among the above five countries. This statistic shows that while India has taken some measures to manage fishery resources, the aquaculture industry has rapidly developed as a major source of export earnings.

3.6.4   Development Trends of Fisheries in India

Due to the limitations of diversity and optional fishing gear, fisheries are faced with many challenges in India, such as an excessive fishing intensity, declining fishery resources, inappropriate development methods, environmental pollution of habitats, water eutrophication, and resource depletion. The Indian government has realized the importance of aquatic ecosystems and their dependence on natural resources, and the central and local governments have enacted and implemented relevant policies, laws, and regulations to promote the sustainable development of fisheries.

In the future, India will continue to attach importance to aquaculture development; increase the productivity of all aquaculture waters; introduce advanced fisheries technology and equipment; improve supporting facilities for deep-sea fisheries; develop deep-sea oceanic fisheries; to conduct training for fishery workers, enhance fisheries culture; to develop marketing techniques; strengthen the management of water leasing, ownership, and fishing areas; and ensure the implementation of responsible and sustainable fisheries activities.

References

Chen XJ, Zhou YQ (2018) Introduction to fishery. China Science Press, Beijing. (in Chinese)
FAO (2018) Fishery and aquaculture statistics 2016. Rome

Chapter 4
An Overview of Major Fishery Disciplines

Xinjun Chen, Yinqi Zhou, Liming Song, Zhihe Wang, Kaijun Wu, and Chen Sun

Abbreviations

3S technology	An abbreviation for Remote Sensing (RS), Geography Information Systems (GIS) and Global Positioning Systems (GPS)
AIS	Automatic identification System
ASFA	Aquatic Sciences and Fisheries Abstracts
AVHRR	Advanced Very High Resolution Radiometer
CZCS	Coastal zone color scanner
DBMS	Database management system
DS	Database system
ES	Expert system
FIS	Fisheries Information System
FiSAT	FAO-ICLARM Stock Assessment Tool
FISHSTATJ	Software for Fishery and Aquaculture Statistical Time Series
G-GIS	Gulf Geographic Information System

X. Chen (✉) · Y. Zhou · L. Song
College of Marine Sciences, Shanghai Ocean University, Lingang New City, Shanghai, China
e-mail: xjchen@shou.edu.cn; yqzhou@shou.edu.cn; lmsong@shou.edu.cn

Z. Wang
College of Food Sciences and Technology, Shanghai Ocean University, Lingang New City, Shanghai, China
e-mail: Zhwang@shou.edu.cn

K. Wu
Library, Shanghai Ocean University, Lingang New City, Shanghai, China
e-mail: kjwu@shou.edu.cn

C. Sun
College of Economics and Management, Shanghai Ocean University, Lingang New City, Shanghai, China
e-mail: Chensun@shou.edu.cn

© Science Press & Springer Nature Singapore Pte Ltd. 2020
X. Chen, Y. Zhou (eds.), *Brief Introduction to Fisheries*,
https://doi.org/10.1007/978-981-15-3336-5_4

131

GIS   Geographic Information System
GnRH   Gonadotropin-Releasing Hormone
GPRS   General Packet Radio Service
GPS   Global positioning system
IP   Internet Protocol
IT   Information technology
LAN   Local area network
LRH-A   Luteinizing hormone-releasing hormone analogues
MEDS   Marine Environmental Data Systems
MGET   Marine geospatial ecology tools
MIS   Management Information System
MPAs   Marine protected areas
NMFS   National Marine Fisheries Service
NOAA   National Oceanic and Atmospheric Administration
NTZs   No-Take Zones
ODBMS   Operational Database Management Systems
PC   Personal computer
RS   Remote Sensing System
UNCED   UN Conference on Environment and Development
USDA   United States Department of Agriculture
VMS   Vessel Monitoring System
WAN   Wide area network

4.1 An Overview of Piscatology

4.1.1 Concept and Research Content of Piscatology

4.1.1.1 Concept

Piscatology is an applied discipline focusing on fishing tools, adaption of harvesting methods, and formation and changing regulation of fishing sites that applies to various species, habitats, quantities, distribution and migration of target fish, and characteristics of natural aquatic environments. Piscatology is a subdiscipline of fisheries science that is closely linked to other disciplines, such as aquatic resource science, oceanography, and meteorology (Chen and Zhou 2018).

4.1.1.2 Research Content

The fishing industry is an important component of fisheries. Fishing is one of the oldest production activities and formed alongside human hunting for natural resources. The fishing industry will permanently endure as an industry in modern fisheries mainly due to its direct capture of fish and other aquatic living resources in nature as well as its economic characteristics.

As a production activity, the fishing industry involves fishing tool design and manufacture; understanding and grasping fishery resources and fishing grounds; support of engineering techniques, such as fishing vessel design and manufacture; freezing and refrigeration of aquatic products; fleet management; and marketing. Therefore, from a disciplinary perspective, scientific and technological support of the fishing industry includes disciplines involving piscatology, fishery resources, and aquatic products processing and refrigeration.

There are three categories under fisheries research: fishing gear science, fishing methodology science, and fisheries oceanography. (1) Fishing gear science focuses on studies of fishing tool design, material properties, and assembly technology. This field can be further divided into "fishing gear materials science," focusing on the fishing gear material and the physical and mechanical properties of related parts; "fishing gear technology," focusing on fishing gear assembly design and technology; "fishing gear mechanics," focusing on fishing gear and the static and hydrodynamic properties of parts; and "fishing gear design science," focusing on fishing gear design and selectivity. (2) Fishing methodology science studies the technology of fish stock exploration, technology of attracting and controlling fish stocks, exploration of central fishing grounds, and adjustment technology during operations and fishing automation. (3) Fisheries oceanography is also known as "subdiscipline of fisheries ground," an applied discipline focusing on the formation and change mechanism of fishing grounds and the changing regulation of fishing conditions. This category involves the relationship between the marine environment and fish behavior, fish detection, and fisheries forecasting.

The research content of the main fishery disciplines is listed below (Chen and Zhou 2018).

1. Fishing gear materials science. This discipline studies and analyzes the structure and performance of fishing gear materials and experimental research methods based on general theory. The components of fishing gear include netting twine, mesh, string, floats and sinkers, and other attachments, of which the most important is fiber. Fishing fiber can be divided into two major categories: plant fiber and synthetic fiber. Before the 1940s, almost all fishing fibers were plant fibers, while after the 1940s, synthetic fibers were widely used in fisheries. Synthetic fibers are characterized by nonperishability, abrasion resistance, high tensile resistance, and low density, characteristics especially suitable for fishing; thus, fishing gear made of synthetic fiber has low resistance, high strength, and a high catch rate.

2. Fishing gear mechanics and fish behaviors. Fishing gear mechanics and fish behaviors are fundamental parts of piscatology. Fishing gear mechanics mainly studies the hydrodynamics of fishing gear components and changes in the shape and resistance of fishing gear under external forces. Before the 1950s, fishing gear theory was represented by Professor Baranov, a Soviet scholar, and Dr. Kenichirotanesen (たうちもりさぶろう), a Japanese scholar, who focused on theoretical studies of fishing gear. In the 1970s, theories were applied to model

tests, and in the 1980s, ponds were built for fishing, facilitating the standardization of fishing gear tests, which provided a theoretical basis for improving fishing gear structure and optimized fishing gear design. Fish behavior mainly focuses on fish responses to external stimuli, including thigmotaxis, galvanotaxis, phototaxis, rheotaxis, chemotaxis, and cybotaxis of fish.

3. Fishing gear and fishing methodology science. Fishing gear and fishing methodology science is the core of piscatology. Fishing gear consists of the tools used to directly harvest aquatic economic animals. According to *The Classification and Nomenclature with Its Code Name of Fishing Gears* (1985), the national fishing gear standard published by the National Bureau of Standards, Chinese modern fishing gear falls under twelve categories, namely, gill nets, surrounding nets, trawls, beach seines, swing nets, lift nets, dip nets, falling gear, traps, lines, rakes and spears, and baskets and pots. A fishing method is a production technology for catching aquatic economic animals with specialized fishing tools for fishing benefits. Fishing methods involve fish stock detection, target fish behavior control, and fishing gear application. Fishing methods includes trapping, roundup, guiding, and anesthesia.

4. Fisheries oceanography. Fisheries oceanography mainly studies the basic principles of fishing ground formation and fish migration. Fishing grounds can be categorized as continental shelf fishing grounds, stream fishing grounds, upwelling fishing grounds, vortex fishing grounds, and reef fishing grounds. Fish migration is closely associated with colonization, an adaptation property to ensure species survival and population increase. Fish migration can be categorized as spawning migration, feeding migration, and overwintering migration.

From the perspective of relationships among disciplines, fishing gear science and fishing methodology science are mainly based on fishing gear mechanics and fish ethology. Fishing gear mechanics is mainly based on theoretical mechanics, material mechanics, fluid mechanics, elasticity mechanics, and structural mechanics. Fish ethology involves fish ichthyology, fish physiology, fish behavior, and many other modern sciences. Fisheries oceanography involves fish physiology, environmental science, meteorology, oceanography, fish ethology, and statistics.

4.1.2 Status and Development Trends of Research on Piscatology

4.1.2.1 Importance of the Fishing Industry and its Sustainable Development

The fishing industry is a major component of global fisheries. Although aquaculture production has increased rapidly over the past decade, fishing production was approximately 95 million tons, accounting for approximately 70% of the total production of 135 million tons. Marine fishing is the major contributor to fishing production. Marine fisheries provide human beings with a large amount of high-quality and low-cost protein, creating considerable economic and social benefits. However, for economic and management reasons, the global fishing capacity has rapidly increased, while overfishing has, in turn, put heavy pressure on fisheries resource conservation, coupled with climate change impacts, prompting the exhaustion of most commercial fishery resources. This harsh reality forces human beings to conduct research on the biological characteristics of fishery resources and fisheries economic laws, explore scientific management methods, seek technical support, and rationally utilize living aquatic resources, thus achieving the sustainable development of marine fisheries.

In the twenty-first century, "resources, environment, and population" and "food security" are hotspot issues for governments, departments, organizations, and human beings all over the world, while fisheries are closely related to these issues and receive attention from different stakeholders. To meet the human need for food and protein and to rationally utilize fishery resources, science and technology should be highlighted to facilitate the understanding of biological and economic laws of fisheries renewable resources, scientific management, and sustainable development.

4.1.2.2 Piscatology and Improvement of Science and Technology

There is historical evidence that science and technology have a broad and profound impact on marine fisheries. For example, the development of the shipbuilding industry and ship motorization has expanded marine fisheries from coastal areas to inshore, offshore, and deep-sea areas. Meanwhile, the structure of the marine fisheries industry and connotations of production modes have profoundly changed. Fishermen used to conduct fishing with vessels powered by manpower or wind from sunrise until sunset in coastal waters, whereas currently, fishermen can conduct fishing far from base ports in deep seas. At present, modern fisheries are represented by deep-sea fishery fleets, which are operated on the basis of large-scale industrial production, specialization, and complex organizational management. Examples include large trawl processing vessels, large, light-luring squid angling vessels, ultralow temperature tuna jiggers, and surrounding net tuna fishing vessels. The production activities of deep-sea fisheries reflect a high degree of science and technology integration, although deep-sea fisheries were not conducted until the 1950s when the science and technology was developed.

The main fishing gear types are stationary (fishing) gear (such as trawls, surrounding nets, jigging, swing nets, baskets, and pots), gill nets, and others. (1) Trawls are the most commonly used fishing gear, a bag-type, drainage netting gear towed by a fishing boat, and fish, shrimp, and other aquatic organisms are isolated and captured through the mesh in the purse. Trawls are a proactive and highly productive fishing gear, with single and double vessels towing a fishing net in the bottom or middle aqueous layer. The net mouth of a large, middle-layer trawl covers an area of nearly 1000 m^2, draining approximately 5 million m^3 of water/h during operation, with catches of more than 10 tonnes/haul. Vessels with large, middle-layer processing trawls are tailored for unrestricted ocean voyages, feasible for conducting year-round, continuous operation at sea. (2) A surrounding net is a large-scale gear type, with a mesh length of 800–1200 m and a height of 100–200 m during operation. During operation, a surrounding net initially forms a cylindrical body in the water. After the bottom wire rope is tightened, the net forms a bowl shape, containing 22 million cubic meters of water. Targets for a surrounding net are mainly clustering fish, such as mackerel and herring, with catches of hundreds of tons per net. (3) Long lines are the main fishing gear for line fisheries, and the main line, which can be up to several kilometers long, can be suspended with sublines with thousands of hooks attached to the lines. Generally speaking, an automatic bait casting machine and angling machine are used in this fishing operation. Tuna long-line fisheries are one of the primary fisheries where the main line is left in a horizontal direction. In addition, as for vertical longlines, such as in squid fishing, many hooks are linked with the main line. When fishing hooks are vertically cast into water to a certain aqueous layer, they have a continuous up-and-down movement. Hooks with an up-and-down movement can lure squid to take the bait. Large fishing boats are often equipped with dozens of automatic squid angling machines, with dozens of 2000-watt fish luring lights, and a light field is formed by the rays through water. Fishing efficiency can be improved by luring squid to form groups by utilizing their light-loving habits. (4) Large gill nets are used to catch fish once a fish is restricted or hooked when moving through the mesh. The net length can reach up to more than 10 km. Fishermen can conduct continuous fishing operations if a gill net is equipped with drift net casting and hauling machines and other specialized machines. Gill nets are categorized into set gill nets and drift nets. Set gill nets are usually fixed by anchors or heavy underwater objects. During operation, set gill nets are fixed at a certain aqueous layer, mostly in shallow water. Drift nets are attached to vessels by fiber ropes, drifting together with the vessel. Gill nets are passive fishing gear. Set nets are further divided into swing nets, traps, falling nets, and drop nets. Fishing gears designed to attract fish to nets according to their habits and characteristics are passive nets (FAO 1957, 1963, 1970, 1988).

With numerous scientific and technological achievements, the fishing industry, as a commercial production activity, is developing quickly. Considering the impact of technology on the fishing industry, three scientific and technological inventions have had a revolutionary effect on modern marine fisheries since the industrial revolution in the eighteenth century, contributing to an unprecedented growth in marine fisheries. These three inventions are "steam and mechanization, artificial and synthetic fibers, and ultrasonic detection equipment," and their contributions are listed as follows (FAO 1957, 1963, 1970, 1988):

Marine fisheries development has contributed to industries such as shipbuilding, electronics, machinery, the chemical industry, refrigeration engineering, and food processing. This long-term and close cooperation has formed an interrelated industrial chain and has enhanced specialization such as fishing vessel construction, fishery electronic and mechanical equipment, aquatic products chemistry and refrigeration, aquatic products processing, and other professional fields, further enhancing marine fisheries. Historical experience shows that proactively introducing advanced scientific and technological achievements to fisheries is an important way of promoting fisheries. In addition to the three major scientific and technological achievements that have contributed to the revolutionary development of marine fisheries, there are many other important scientific and technological achievements that have had positive effects on marine fisheries and the fishing industry, such as the introduction of large-scale, factory chemical trawlers and plate freezers. The large-scale, double-deck stern ramp trawl processing vessel is an important invention for modern fisheries development, because it integrates fishing, processing and production, refrigeration and preservation, and transportation together in one vessel that serves as a one-stop marine processing plant. Vessels are designed for unrestricted ocean voyages, and endurance is greatly extended; thus, we can travel around the world in various sea areas to conduct year-round operation. Refueled and unloaded at sea, fishing vessels can carry out continuous operation at sea for a long time. These activities, in turn, require the support of well-structured vessels, advanced communication technology, and advanced marine loading and unloading methods to adapt to various climate and sea conditions and ensure safe operation.

In fisheries production and fisheries science research, it is important to observe fishing gear status during operation and the behavior and behavioral reaction of fish; however, these aspects are not easy to observe in a vast and deep ocean. As for classifying observation technology, it can be divided into in situ observation (aerial observation and underwater observation) and laboratory observation.

To improve fishing and production efficiency, selective fishing is applied to achieve the sustainable utilization of fishery resources, and fishing production tools must be adapted to fish behavioral reactions and the production environment. Fishing gear mechanics and fish ethology are two disciplines playing important roles in conducting research on ecofriendly fishing gear and methods. (1) Fishing gear mechanics conducts research on the mechanism of tension distribution and spatial shape of gear components, such as mesh, netting twine and ropes, under current action and various loads, and on the formation of corresponding hydrodynamic force and resistance, thus providing data for fishing gear design. Fishing gear mechanics mainly focuses on the following research areas: fishing gear design and calculation; model test principles and methods; mathematical mechanical modeling of fishing gear shape and operating performance; computer simulation and real-time control; trawl expansion equipment design; spatial position and motion of fishing gear; and fishing gear operation system monitoring and data processing. (2) Fish ethology and selective fishing methods are also important. During fisheries production, whether in capture or aquaculture, it is important to understand fish behavioral habits. As for aquaculture, it is important to understand fish behavioral habits, such as feeding and reproduction. As for those whose work involves fisheries resource conservation, it is important to learn the migration laws of fish stocks. As for capture, it is important to have adequate knowledge on the reaction of fish species or fish stocks to fishing gear and their behavioral principles during fishing production, thus applying selective fishing to protect resources and improve production efficiency.

4.1.2.3 Concept of Balanced Fishing and Its Practice

The term "balanced fishing" originates from "balancing/balanced exploitation," "balanced harvest," and "balanced fishing." Fishing involves fishery resources, fishing grounds, fishing methods; thus, there are different interpretations of "balanced fishing" in different research areas. "Balancing/balanced exploitation" and "balanced harvest" are interpreted from an ecological perspective, focusing on the development and utilization objects of marine ecological resources, prioritizing developing and harvesting biological resources. "Balanced fishing" is interpreted from the perspective of fishing gear, emphasizing the methods of developing and utilizing biological resources, which refers to the use of appropriate fishing gear and taking advantage of all fishery harvests (including bycatch and miscatch), rather than discarding fish after death or causing death after discard, both of which lead to considerable resource waste.

However, "balanced fishing" refers to a management strategy of diffusing fishing stress (mortality) to all trophic levels to maintain healthy trophic relationships among fish of different species and size (Zhou 2013). Balanced fishing is usually denoted with a nutrition pyramid, which indicates that fishing activities should be carried out at different trophic levels to remain positively proportional to the corresponding productivity.

Fisheries activities are often characterized by selectivity, because fishing targets are usually those species and/or individuals of a certain size that contribute to maximum economic returns. Moreover, gear types are all characterized by selectivity. In spite of different selectivity, the core of any gear is its technical characteristics and configuration. Selectivity appears at different levels. For example, specific fishing gear is used in fishing to catch fish of a targeted species and size; fishing is conducted at the specific fishing grounds where fish of a targeted species and size are available. Selective fishing practices may result in composition changes in communities or ecosystems where changes may occur in individual sizes and species. Some fishing activities are aimed at particular trophic level species (such as krill, small pelagic fish, or top predators); therefore, these activities directly take away an ecosystem component without considering the staircase effect on dependent species, which can also be seen as a selective fishing form at the ecosystem level. There is evidence that the more diffuse targeted species and individual sizes, the higher is the harvest. Instead, if fishing activities unevenly affect different trophic levels, these activities may then change ecosystem structure, resulting in production decline (Garcia 2011;Smith et al. 2011).

Over the past few decades, it has already been realized that the trophic relationships in a particular ecosystem should be considered in fisheries policy making; otherwise, as confirmed by research findings, fishing activities will negatively impact the structure and operation of an aquatic ecosystem (Garcia 2011; Smith et al. 2011).

In the early 1970s, the growing interest in krill fishing in the Southern Ocean attracted concern from the international community, because krill plays a key role in the Antarctic food chain; besides, krill fishing may negatively impact predatory species. With increasing demand from the international market for low trophic level species, such as krill, sardines, anchovies, and herring, fishing for these species has recently triggered concern. These species not only play important roles in ensuring food security but are also important animal feed (e.g., in aquaculture). Moreover, these species also play key ecological roles in the conversion production process from plankton to relatively larger predatory fish, marine mammals, and seabirds. To leave enough food for marine predators, a much more conservative and sustainable catch rate has been proposed, which is far more below the maximum sustainable yield.

As for balanced fishing, another concern is shrimp fishing in tropical regions. Various bottom trawls with a minimum mesh size of the purse (including beam trawls) are used in shrimp fishing; therefore, these trawls have low selectivity, and they are regarded as harmful as they often result in large bycatches of species that are more easily damaged than shrimp. Fishing gear for shrimp fishing could have a greater impact on associated species, because these species tend to have lower productivity (i.e., low fertility rate and slow growth rate). Compared with shrimp, these species have longer life spans (i.e., low replacement rate); thus, they are more likely to be damaged. Shrimp fishing may also lead to changes in the fish community structure and may negatively impact other species that are targeted species in other fishing activities.

"Balanced fishing" has recently been used to assess the impact of fishing on relatively larger-sized individual fish and species (usually those in a relatively higher level of the nutrition pyramid and of relatively high economic value). Traditional fisheries management strategies are usually based on selective fishing. A case in point for a traditional strategy is the requirement for a minimum mesh size (with a view to protecting fish that have not reached first maturity), which may alter the food chain structure, resulting in low productivity and declining aquatic ecosystem recovery ability and finally leading to smaller size of mature and premature individuals due to phenotypic changes and fast growth. In addition, balanced fishing requires strict supervision, which requires manpower and financial support; thus, it is a difficult and high-cost endeavor to conduct supervision. Therefore, it is suggested that selective fishing with an individual size as the fishing standard should be abandoned, which is believed to realize balanced fishing, maintain the ecosystem structure, and reduce transaction costs for supervision. This suggestion has triggered a debate, and it has been seen as legislation, which is likely to subvert current fisheries legislation of most countries in the world.

The idea of better maintaining ecosystem structure and operation through a more balanced fishing strategy literally sounds very reasonable and scientific (Zhou 2013). It is widely acknowledged that a more global management approach, rather than the single-species management approach, should be adopted. It is also widely recognized that fishing brings about "collateral damage" to aquatic ecosystems caused by fishing activities. Moreover, determining a low-cost and practical fisheries management strategy and approach is more important. On the one hand, it is important to establish an ideal fishing approach; on the other hand, it is also essential to consider relevant social and economic implications and restricting factors.

Currently, it is widely believed that it is very important to adopt a "balanced" approach to developing and utilizing marine ecosystems to enhance the development of ecosystem-based fisheries management and a fisheries ecosystem–based approach. Meanwhile, the importance of maintaining biomass at each trophic level and maintaining the abundance of individuals with different sizes at different trophic levels are also widely recognized. Many discussions concerning the above issues have been carried out. The main challenge now is to convert these ideas into fisheries management practices.

4.2 An Overview of Fisheries Resources Science

4.2.1 Concept and Research Content of Fisheries Resources Science

Fisheries resources science is one of the main fishery disciplines. This field refers to scientific research on the natural life history (breeding, feeding, growth, and migration) of a fishable population, laws of population quantity change, estimated quantity of resources and allowable catches, and fisheries resource management and conservation measures. This research aims to provide a scientific basis for reasonable fisheries production and scientific fisheries resource management. With the development of this discipline, the connotation of fishery resources is constantly extended; therefore, fisheries resources science is a science currently focusing on the following elements: the natural life history of a population, including fish and other species, concerning breeding, feeding, growth, and migration; the law of quantity changes in populations, including fish and other species; the estimated quantity of resources and allowable catches; the law of population quantity changes and uncertainty under different management strategies; the law of the relationship between resource development and utilization of fish species and social and economic development and the optimum distribution of resources; and fisheries resource management and conservation measures, to provide a scientific basis for reasonable fisheries production and scientific fisheries resource management.

Fisheries resources science usually includes fisheries biology, fisheries resource assessment and management, and enhancement of fishery resources. These sub-disciplines are described below (Chen 2014):

1. Fisheries biology. Fisheries biology studies aquatic economic floral and faunal life history, population characteristics, age structure, growth, mortality, feeding, breeding, population supplement, and characteristics of quantity changes, which are the main research areas of fisheries science. Fisheries biological characteristics are the basis of resource assessment, fishing condition forecasting, resource management, and enhancement, focusing on not only biological issues related to

fishing but also on scientific issues that involve economic, social, and management issues related to fisheries.

2. Fisheries resource assessment and management. Fisheries resource assessment and management studies population dynamics and quantity changes in fishery organisms (mainly economic fish), which falls under fisheries population ecology, and is the core of fisheries resources science. With knowledge on the biological characteristics of fish stocks as the basis and the ideal stocks as the hypothetical premise, fisheries resource assessment means to simulate and predict the past and future status of fish stocks to provide a scientific basis for policy making for fisheries resource conservation, by establishing mathematical models, describing and assessing the composition structure of stocks, quantity and changes of resources, assessing the impact of fishing, the environment and other factors on fish populations, and understanding the characteristics and laws of resources quantity change. The main purpose of fisheries resource assessment is to achieve the sustainable utilization of fishery resources under given management goals through quantitative analysis of fishery resources, fully considering uncertainties in fishery resources, seeking the most optimal measures for fisheries resources development and management, providing a scientific basis for such measures, and conducting risk analysis on these measures.

3. Enhancement of fishery resources. The enhancement of fishery resources consists of cultivating and proliferating fishery resources, improving the aquatic environment, and maintaining an ecological balance by artificially releasing fish, shrimp, shellfish, algae, and other aquatic organisms' larvae (or adults, eggs, etc.) into natural waters to increase resource quality and to improve and optimize fishery resources population structures. Broadly speaking, the enhancement of fishery resources also involves indirect means such as improving the aquatic ecological environment, constructing some structures (such as structures for attaching eggs, artificial reefs, etc.) and releasing some wild stocks into specific waters. Specific biological resource populations in specific waters would decline and be depleted if external conditions change, especially when the impact of human activities (such as overfishing, water pollution, damming rivers and disconnecting lakes, etc.) has exceeded the environment's capability to restore itself. Then, various measures are to be taken to increase its supplementation to maintain the optimum resource quantity. Various biological populations in aquatic ecosystems are for human use; however, some of them directly or indirectly endanger aquatic resources. To integrate the ecological processes in an aquatic ecosystem with economic organism production to improve the efficiency of energy conversion from primary production to economic organism resource production, it is necessary to introduce new feed organisms and even important economic organisms according to the aquatic ecological conditions, to change the original biological community composition, and to increase the species variety and resource quantity.

4.2.2 Status and Development Trends of Fisheries Resources Science

4.2.2.1 Research Status of Fishery Resources Biology

Fishery resources biology, a subdiscipline of biology, is a natural science, studying fish resources and the population ecology of other aquatic economic animals. Usually, fishery resources biology means "research on the population composition of aquatic organisms, such as fish, as well as research on the life history and characteristics of fish populations, such as age composition, growth characteristics, sexual maturity, breeding habits, feeding, and migratory distribution at each stage of a fishery organism's life cycle." The research on fishery resources biology dates back to the year of 1566. Robert Hooke observed the structure of fish scales with the then newly emerged microscope. Ever since then, identification of fish scales has long been a major research topic for fishery resources biology.

With the advancement of research methods and science and technology, the overlap between disciplines has been enhanced, and microchemistry and microstructure technology have been applied. Researchers can assess fish age using tiny otoliths and other materials, and the "daily growth increment" can be tracked daily through otoliths. By observing the distribution of these increments, researchers can better analyze fish growth in early development during a day, week, or season. Meanwhile, researchers can also analyze the microelements of hard tissue, such as otoliths, to study the relationships between content changes and various growth and marine environmental factors, such as changes in water temperature, to speculate on fish life history and species composition.

At the same time, researchers have a better understanding of all aspects of fishery resources biology, and the overlap and integration among molecular biology, physiology, bioenergetics, and environmental science has promoted the development of fishery resources biology, formed some emerging disciplines, and developed some research areas, such as fish molecular phylogeography, fish breeding strategies, environmental factors for fish breeding, and the impact of climate change on fish stocks.

Fish populations are an important part of fishery resources biology, as well as the basis and difficulty of this study; therefore, technology relating to population identification and discrimination has developed rapidly. Recent major academic classics are (1) *Geometric Morphometrics for Biologists*, co-edited by Miriam Leah Zelditch, Donald L. Swiderski and H. David Sheets (2012), a monograph that systematically introduces geometry morphology and its application in fish population identification and (2) *Stock identification methods: applications in Fisheries Science*, co-edited by Steven X. Cardrin, Lisa A. Kerr and Stefano Marianl (2014), a monograph that systematically describes the present study methods of population identification in the world, especially the development of some new technology and methods, such as the latest research results in molecular biology, microchemistry, image recognition technology, and tag and release technology.

Moreover, the overlap among multiple disciplines promotes population identification technology and enriches the research content and technology systems of fishery resources biology.

4.2.2.2 Development Status of Fisheries Resource Assessment and Management

Fisheries resource assessment is conducted on the basis of scientific research and fishing data. By utilizing fisheries resource assessment models, fisheries resource assessment estimates parameters related to fisheries and stocks to trace back to a population and fishing history, assess fishing activities and the impact of fisheries management on resources, forecast the development trends of fishery resources, and analyze risks. Therefore, fisheries resource assessment is the basis for the scientific management of fishery resources.

With enhanced computing capabilities and driven by multiple disciplines, fisheries resource assessment models have developed rapidly over the past 30 years. Researchers are continuously expanding assessment models to take advantage of a variety of data sources to more vividly describe population dynamics. The parameter estimation method has become more diverse, parameter estimation uncertainty quantification has been improved, and the effect estimation of management strategy has been more comprehensive. Fisheries resource assessment models have been increasingly complex and diverse; thus, it has been relatively difficult to decide on a model, and inappropriate model use may result in fisheries resource management failure.

Usually, most fisheries resource assessment models consist of four submodels (Guan et al. 2013): (1) population dynamics model, a model that simulates a population's dynamic changes in accordance with the life history characteristics of this population and fishing process; (2) observation model, a data prediction model (such as catches and resource indices); (3) objective function, defining an objective function through observing variate error structure hypothesis and prior information and estimating parameters through minimizing or maximizing the objective function; (4) projection model, a model that analyzes a population's dynamic changes for a certain period to evaluate the management effect and risk through parameter values and the related management control rules. Different assessment models vary in a population's dynamic simulation, observation model construction, parameterization method, and parameter estimation method.

The main fisheries resource assessment models include the single-species fisheries resource assessment models, multispecies fisheries resource assessment models, and ecosystem-based fisheries resource assessment models. At present, the single-species model is still the major model for fisheries resource assessment and management. However, with the development of computing capabilities and development of related software, the single-species model is becoming increasingly diverse and complex. Due to the various assumption and data requirements of each model, differences in the evaluation effect of different models do not necessarily indicate the pros and cons of a model. Therefore, much attention should go toward evaluation targets and characteristics of the required data. By summarizing experiences and lessons learned, researchers can improve the capability to use assessment models and methods and further improve a model to provide appropriate advice on fisheries resource assessment and management using the most appropriate model. Meanwhile, both multispecies and ecosystem-based fisheries resource assessment models should also be encouraged to provide data, knowledge, or theories to improve the assessment and management quality of single-species assessment models. Combined with the uncertainties concerning fisheries data and assumptions, these models are capable of conducting evaluation on management strategies, and in doing so, management risks can be avoided.

As ecosystem-based fisheries resource management is developing into the most popular fisheries management method, adequate knowledge of interspecies relationships and the impacts of the environment, climate change, and human activities on fisheries ecosystems are important research components on future fisheries resource assessment models, which are believed to be the basis for developing ecosystem-based fisheries resource management. We can determine the initial life stage of fish and its influence on recruitment through a combination of marine physics and biology; we can analyze the impact of marine environmental changes on a population's spatial distribution through establishing habitat models and utilizing marine observational data, such as data from marine remote sensing and assimilation data from physical marine models; we can learn or predict energy conversion processes and trophic relationships, such as predation and competition among stocks at different trophic levels in an ecosystem, through establishing quantity and energy balance models based on the food web; the combination of physics, biology, and fishing will further promote the development of ecosystem-based fisheries resource assessment models. Ecosystem-based models will be a trend in the development of fisheries resource assessment models. With more multidisciplinary interactions and increasingly richer observational data, ecosystem-based models will be widely applied in fisheries resource assessment and management.

4.3 An Overview of the Science of Aquaculture and Enhancement

4.3.1 Concept and Research Content of the Science of Aquaculture and Enhancement

4.3.1.1 Concept

The industry of aquaculture and enhancement utilizes appropriate inland waters, shallow seas, and mud flats for aquaculture and enhancement (including enhancement by artificially releasing seed into natural waters, aquaculture environmental protection, and natural enhancement) (Chen and Zhou 2018).

The science of aquaculture and enhancement focuses on the biological characteristics of seawater and freshwater economic aquatic products and their relationship with the ecological environment of aquaculture waters. Based on the ecology, physiology, ontogeny, and population growth of aquaculture species, the research on aquaculture and enhancement are artificial breeding, seed cultivation, and aquaculture and enhancement technology under controlled conditions, with environmental protection, suitable aquaculture waters, and engineering facilities as preconditions (Chen and Zhou 2018).

4.3.1.2 Research Content

Taking a look into the developmental history of fisheries scientific technology and its impact on aquaculture and enhancement and the developmental history of the science of aquaculture and enhancement, the research content of aquaculture and enhancement mainly focuses on water environment control, genetic breeding, nutrition, and feed. Water quality control in aquaculture waters and water ecosystem restoration and protection form the foundation for ensuring aquatic products quality and safety and sustainable aquaculture development. Breeding, seed cultivation, and feed are the material conditions for fish culture. The aquatic environment, breeding, and feed (commonly known as "water, breeding, and bait"), three basic aspects of aquaculture, have become the core of higher education and research in aquaculture science (Chen and Zhou 2018).

Aquaculture Water Environment

An aquaculture water environment is an aquaculture ecosystem, which consists of three parts: consumers (aquatic animals), decomposers (aquatic microorganisms), and producers (aquatic plants). An aquaculture ecosystem is driven by energy flow and the matter cycle. Aquaculture technology can be divided into two systems, that is, traditional and ecological aquaculture ecosystems. (1) The traditional aquaculture ecosystem in traditional aquaculture is extremely unbalanced in that its production technology one-sidedly prioritizes consumers, ignoring decomposers and producers. Traditional aquaculture gives rise to such problems as eutrophication, serious disease, and the use of a large number of drugs to treat disease, which brings about imbalance with the external environment and microecological imbalance in organisms. (2) The ecological aquaculture ecosystem in ecological aquaculture is healthier in that there are no significant adverse environmental effects for a long time. Ecological aquaculture is aimed at establishing and managing a healthy aquaculture ecosystem, which can, in light of ecological principles, be ecologically self-sustainable with low input and economically feasible. Ecological aquaculture prioritizes maintaining and improving the ecological dynamic balance in an internal ecosystem, arranges for a reasonable aquaculture structure and product layout, strives to improve the utilization rate of solar energy, and promotes recycling and reuse of materials in a system to reduce the utilization of fuel, fertilizer, feed, and other raw materials as much as possible. The goal is to gain more fishery products and attain favorable outcomes by equally prioritizing production development, ecological and environmental protection, energy recycling, and economic benefits. Only by using ecological aquaculture techniques can we fundamentally solve the disease problems of aquaculture waters.

The goal of ecological aquaculture is to repair the ecological aquaculture water environment, with emphasis on preventing eutrophication in aquaculture waters and maintaining "two balances" (external environment of aquaculture waters and internal environment of cultured species), and to sustainably develop aquaculture waters. There are three ways to repair the aquatic environment: physical remediation, chemical remediation, and bioremediation.

Aquatic Products Seed Breeding

The successful artificial breeding of Chinese carp in freshwater is a major achievement in the history of China's aquatic science. This success ends the passive situation of our long-term reliance on fishing for natural fry in rivers; instead, we can manually control artificial breeding with plans to produce fingerlings. This process has promoted China's aquaculture and enhancement industry into a new historical period, while conserving fish resources in Chinese rivers. Moreover, the basic principles and techniques of Chinese carp artificial breeding shed light on artificial breeding of other fish in sea or freshwater.

In recent years, research on the reproduction and physiology of marine cultured fish has initially thrown light on the hormonal regulation mechanisms of gonadal development and gamete maturity. Inducing hormones, such as by injecting or implanting the highly active, sustained-release agent of Gonadotropin-Releasing Hormone (GnRH), can promote gonadal development of marine cultured fish, which would then easily ovulate and spawn under artificial breeding conditions. Moreover, by strengthening nutrition, the sperm and egg quality can be improved, the reproductive capacity of parent fish enhanced, and the survival rate of offspring increased. Furthermore, through environmental regulation, especially regulation of the water temperature and photoperiod, marine cultured fish can reach gonadal maturity and spawn throughout the year. At present, more than 40 species of marine fish have been successfully reproduced artificially. The number of large yellow croaker fingerlings is over 100 million. The number of larvae of barracuda, red snapper, amoycroaker, crescent sweetlips, Japanese sea perch, and American red fish is over 10 million. The number of fingerlings of black snapper, skewband grunt, banded grunter, and bastard halibut is over 1 million. However, many fish fry cannot be reproduced on a large scale; therefore, the problem of artificially breeding marine fish fry has only been partially solved to date.

Nutrition and Feed

Feed is an important element in aquaculture. To provide the necessary foundation material for normal growth and good production of cultured animals, the cost of feed accounts for the highest percentage of aquaculture production cost, generally as much as 50%–70%. Therefore, improving animal feed utilization has a great influence on the economic benefits of aquaculture.

As for the application of fish feed in aquaculture, much attention was previously paid to animal growth, feed utilization, and feed costs, with little attention paid to its environmental impact. In aquaculture practices, although artificial feed can ensure the required nutrients for the growth of cultured animals, nitrogen and phosphorus in feed may contaminate the aquatic environment. Nitrogen and phosphorus in feed cause environmental pollution in three aspects. First, nitrogen and phosphorus from the feed that cannot be fully ingested by animals after absorption is discharged into the water in the form of feces, which brings about pollution. Second, absorbed amino acids can not only be used to synthesize protein but also serve as an energy source for oxidization, during which nitrogen-containing compounds as its metabolites are excreted outside of the body through gills and urine. Third, with excessive feeding, residual bait in the water is decomposed by bottom microorganisms, and nitrogen and phosphorus in them is released in the free form. Discharging a large amount of nitrogen and phosphorous from feed into the aquatic environment results in eutrophication, plankton blooms, loss of the aquatic environment ecological balance, decreased pH, reduced dissolved oxygen, increased ammonia nitrogen, and an increased number of planktonic bacteria, which multiply the disease occurrence rate in aquatic animals.

In recent years, to protect the aquatic environment, much attention has been paid to environment-friendly aquaculture feed. Measures have been taken to produce environment-friendly feed. First, the feed formulation has been set to a reasonable protein-calorie ratio to minimize the utilization rate of feed protein as an energy source and to reduce the excretion of nitrogen-containing compounds, the protein decomposition product. Second, feed protein has been kept at a reasonable level, the nutritional value of the feed protein has improved, and the feed protein level has been based on the nutritional needs of the aquaculture animals. Moreover, the feed protein digestibility should be improved to maintain a balance of essential amino acids in the feed protein and increase feed protein availability. Third, the utilization of phosphorus in feed has been improved. In addition, perfect formulation is not enough for good fish feed; excellent feed-processing technology and favorable feeding techniques are necessary as well. In doing so, feed utilization can be enhanced, and feed pollution to the environment can be reduced.

4.3.2 Status and Development Trends of Global Aquaculture

4.3.2.1 An Overview of Aquaculture Development in China

China has a long history of the aquaculture industry, but little attention was paid to aquaculture science as a discipline until the founding of the Peoples of Republic of China in 1949. Since the founding of New China, the main achievements of aquaculture science in China consist of the following aspects (Chen and Zhou 2018):

1. Researchers led by Zhong Lin, an expert in fish breeding science, in May 1958, initially broke through the technical difficulties for artificial breeding of chub and bighead carp in ponds by successfully hatching fish fry. This achievement played an important role in fish fry production in China and even in the development of the global aquaculture industry, because it laid the foundation for a large, local supply of fish fry. Since then, through the same principles and methods, Chinese fisheries researchers have broken through the difficulties in artificially breeding dozens of cultured fish and rare fish, such as grass carp, black carp, mud carp, blunt-snout bream, catfish, Chinese sturgeon, longsnout catfish, bass, bastard halibut, and large yellow croaker, which facilitated the development of polyculture, intercropping, and aquaculture production. This research breakthrough laid a solid foundation for the great development of the aquaculture industry in China.
2. In 1958, through summarizing fishermen's rich culturing experience, eight key techniques for aquaculture were summed up as "water, species, bait, mix, density, rotation, protection, and management," referred to as the "eight key aquaculture techniques," which have established a complete technical system of aquaculture. Chinese fisheries researchers have had in-depth studies on the theory of high yield and efficiency in various waters, culturing methods, and aquaculture regulations for more than five decades, and they finally succeeded in summarizing the ecological aquaculture technique system of high productivity and efficiency in different waters. Additionally, this system has been expanded for large-scale application in a short time in China and has achieved great social, economic, and ecological benefits.
3. Intensive research has also been conducted on the mechanism of oxytocics and reproductive physiology of fish. We are the first in the world to apply oxytocics to artificial fish breeding at a large scale and the first to synthesize luteinizing hormone–releasing hormone analogs (LRH-A), thereby enhancing the effect of induced fish spawning and productivity of artificial fish breeding.
4. Through methods such as introduction and domestication, genetic breeding, and bioengineering technology, a large number of fish are listed as optional species for aquaculture. Since the 1990s, more fish species with a good reputation, special characteristics, and of high quality have been listed as optional species for aquaculture. The main new breeding objects are listed as follows: Chinese sturgeon, amur sturgeon, hybrid sturgeon, Russian sturgeon, rainbow trout, silver fish, eel, lotus carp, Jian carp, three hybrid carp, Furong carp, altered silver carp,

Pengzecrucian carp, Qihecrucian carp, carmine fish, rohu, largemouth catfish, Clarias leather, longsnout catfish, channel catfish, rice-field eel, Chinese perch, bass, black bass, striped bass, Nile tilapia, Bluetilapia, tilapia, fugu, large yellow croaker, red snapper, bastard halibut, grouper, and Chinese black sleeper, among others. Introducing these new aquaculture species enhances the aquaculture industry by improving yield and efficiency.

5. Much research has been conducted to look into the physiological needs of several major aquaculture species in China and to explore their need for protein, necessary amino acids, fats, carbohydrates, vitamins, and various minerals, thus providing a theoretical basis for fish feed production. In recent years, efforts have been made to combine fish feed with our traditional, comprehensive fish aquaculture method, a technology that has accelerated fish growth, improved feed efficiency, and brought economic benefits.

6. Research on aquaculture ecosystems has been enhanced. Through taking ecological aquaculture measures, such as creating underwater forests, applying microecologics, and reforming aquaculture modes, the aquatic environment is restored and protected, the energy flow of a water body is maintained, and the matter cycle is balanced. Adopting healthy aquaculture technology not only improves the quality and safety of aquatic products but also promotes the development of water-protected fisheries and low-carbon fisheries, thus maintaining the sustainable development of aquaculture.

7. Long-term research has been conducted on the prevention and control of common and frequently occurring diseases in the main aquaculture species, and these fish diseases are largely under control. In recent years, the focus on disease prevention and control has shifted to improving the ecological conditions of aquaculture fish by promoting ecological disease prevention and implementing healthy aquaculture, thus preventing disease in aquaculture methods, and great progress has been made.

8. There has been significant development of fishery facilities that focus on the production of fingerlings with a good reputation, special characteristics, and of high quality. Since the 1980s, China's aquaculture industry of aquatic products with a good reputation, special characteristics, and of high quality has taken off, and the need for fingerlings has also surged. However, aquatic fingerlings with a good reputation and special characteristics have poor adaptability to an unfavorable environment; thus, good ecological conditions are a prerequisite. Therefore, greenhouse breeding through artificially controlling the microclimate has become popular. Greehouse aquaculture facilities generally include the following systems: oxytocin systems, series nursery pond systems, water treatment systems, feed and live bait supply systems, heating supply and thermal insulation systems, inflatable oxygen aerator systems, power supply systems, and environmental monitoring and control systems.

9. Preliminary service systems have been established for the aquaculture industry, such as fingerling production and marketing systems (including proto-species farms, good breeding farms and aquafarms), fisheries technology extension service systems, feed supply systems, and service systems for products

circulation. Many practices have proved that establishing these service systems can ensure the smooth and healthy development of the aquaculture industry in China.

4.3.2.2 Status and Development Trends of Global Aquaculture

At present, global fisheries are facing two major problems: how to restore, protect, and sustainably use natural fishery resources and how to ensure the sustainable development of aquaculture. Compared with other important fisheries nations, China is the only fisheries supercountry whose aquaculture production surpasses fishing production. Although China has taken the lead in a number of aquaculture technologies, there are significant gaps in some areas, such as scientific research, management, and information technology, when compared with other developed countries. Therefore, it is necessary to understand and grasp the development trends and directions of the global aquaculture industry and absorb successful foreign experiences to speed up modern fisheries development. The current development trends of global aquaculture can be summed up in three aspects (Chen and Zhou 2018):

1. More attention is paid to aquatic living resource conservation and ecological and environmental protection.

   As human beings accelerate the development and utilization of aquatic living resources and waters, more attention is paid to aquatic living resource conservation and aquatic ecological environmental protection. First, inshore fishery resources enhancement is treated as an important measure for fisheries resource conservation, and artificial release and economic and ecological benefits are evaluated. The FAO has proposed the concept of "responsible aquaculture and enhancement," which calls for evaluating the potential fishery impact on biodiversity and possible alternatives to artificial release, based on resource status and the environment, to implement responsible fisheries resource enhancement. Second, priority should go to the environmental and ecological benefits in shallow sea development and utilization. Before fisheries operate, evaluations should be conducted on the environmental capacity, maximum allowable release capacity, role of releasing a population in ecosystems, aquaculture pollution, and potential dangers that ecological invasions may cause. Some developed countries attach great importance to the research on reservoir limnology, the impact of water conservancy projects on the environment, and research implications. Third, in 1992, "*Agenda 21*" was adopted at the UN Conference on Environment and Development (UNCED), listing protections of various oceans and seas, including enclosed and semienclosed seas and coastal areas, conservation and exploitation of marine living resources as priority issues. Developed countries not only strictly control the discharge of industrial and domestic wastewater but also restrict development of the aquaculture industry, by setting standards for aquaculture wastewater discharge, fisheries drug utilization, special aquatic products circulation, and aquatic wildlife protection, which have formed a series of legal systems.

Much improvement has been made in ecological and environmental restoration and protection of fishing grounds, establishment of aquaculture farms, wastewater treatment, and reduction of pollutant diffusion or accumulation.

2. Efforts have been made to develop and promote efficient intensive aquaculture techniques.

Efficient intensive aquaculture techniques shed light on the trends of modern aquaculture development. Much progress has been made technologically. Deepsea and storm-proof cage aquaculture and industrial aquaculture are developing quickly, and with gradually maturing technologies, more aquaculture species are introduced and the aquaculture field and scope are also expanded. Japan is the first country to begin marine cage aquaculture with high-value fish as the primary aquaculture species. Almost the entire production process can be conducted in cages, including parent fish spawning, fingerling breeding, commercial fish culture, and bait culture; meanwhile, cage aquaculture is also applied to offshore areas. Over the past decade, Norway, Finland, France, and Germany have committed to researching and developing large-scale marine engineering structural cages and aquaculture engineering vessels. These cages have various properties, such as being lightweight, age resistant, easy to install, equipped with an automatic feeding apparatus and monitoring and management apparatus, and capable of withstanding 12-m high waves. Meanwhile, techniques involving solar energy, wind energy, wave energy, tidal energy, and sound and light inducement have been applied to cage aquaculture. Currently, cage aquaculture systems are developing with the trends of being antiwave, automated, and offshore, which has promising prospects. Industrial aquaculture is an intensive, land-based aquaculture method utilizing modern industrial technology and equipment. This method is characterized by a high stocking density and a water- and space-saving, controlled environment, free of seasonal restrictions. This technique has been promoted as a development trend in some countries, thus receiving policy, legislative, and financial support. Countries such as Japan, the USA, Denmark, Norway, Germany, and the United Kingdom have accumulated a large amount of experience and advanced technologies in intensive aquaculture. Examples include the warm running water system invented by the UK Handston Power Station, the biological packet filtering system adopted in Germany, the industrial breeding system for Atlantic salmon in Norway, and the breeding farm for white shrimp in Arizona in the USA. At present, the main form of industrial aquaculture is an enclosed farm with circulating water and diverse aquaculture species (mainly high-quality fish, such as shrimp and shellfish).

3. Modern technology and management have been introduced into aquaculture.

Norwegian Atlantic salmon aquaculture is a case in point for modern aquaculture technology and management. Atlantic salmon aquaculture is conducted in Australia and many other countries in Europe and North America, where the industry is extremely competitive. However, the Norwegian Atlantic salmon industry has maintained rapid growth, with its production ranking first in the world for many years, and this industry has become the second largest pillar industry in Norway. Its success mainly relies on two points: strict governmental management and advanced technological systems.

As for management, in light of the principles and concepts of environmental protection, scientific planning, and farm and products control, Norwegian government departments have strictly implemented an aquaculture licensing system and have established detailed and demanding requirements for the site selection of aquaculture farms, stocking density, professional and management training for practitioners, and prevention of fish disease and pollution. Norway has established an excellent global marketing network, expanding international markets through government funding. As for the technology, much progress has been made in Norway. The aquaculture environment has been significantly improved through transforming cages from large to extralarge, with the cage perimeter increased from the previous 50 m to the present 120 m. Standardization of fingerling quality and feed has been established; thus, fingerlings with fast growth, good adversity resistance, and high disease resistance are chosen as the main aquaculture species. Currently, in Norway, 80% of the production comes from the aquaculture oriented to a single, high-quality species, and feed formulation has also been greatly improved by balancing feed nutrition. Feeding can be precisely manipulated by a computer, with automatic feeding at a precise time, precise site, and precise quantity. In accordance with fish growth, appetite, water temperature, climate change, and remaining bait quantity, the feeding quantity can be automatically adjusted through sonar, television cameras, and the remaining feed collection system; besides, the daily feeding time, site, and quantity can be automatically recorded as well. Vaccines have also been developed and promoted. Four common vaccines have been widely produced. Vaccines can be simultaneously injected; life-long immunity is feasible with a single dose. Moreover, the general application of vaccines not only controls disease and reduces the use of antibiotics but also strongly ensures product quality and safety.

4.4 An Overview of the Science of Aquatic Products Processing and Storage Engineering

4.4.1 Concept and Research Content

The preservation and processing of aquatic products is a discipline focused on processing, utilizing, and maintaining the quality of aquatic products, such as fish, shrimp, shellfish, and algae. Generally speaking, this discipline consists of three professional subdisciplines: preservation, processing, and the comprehensive utilization of aquatic products. Specifically, this field involves cryogenic preservation, curing, dry-curing, smoking, canned food processing, frozen food processing, minced fillet products processing, simulated food processing, feed, other food with aquatic products ingredients, medicine, and chemical products. The basic disciplines involve aquatic animals and plants, analytical chemistry, physical chemistry, biochemistry, chemical engineering, thermodynamics, mechanics, microbiology, food chemistry, food nutrition, and food hygiene. The preservation and processing of aquatic products is an interdisciplinary applied science with high comprehensiveness, a subdiscipline of aquatic science and technology as well as an important subdiscipline of food science and engineering.

The processing of aquatic products has a long history, and there are many processing methods, which can be generally divided into two categories, that is, traditional processing and modern processing. Traditional processing mainly involves traditional crafts, such as curing, dry-curing, smoking, pickling, and natural fermentation. With the rapid economic development in China, continuous progress of science and technology, and introduction of advanced equipment, the methods and means of processing have changed completely, and the technology and added value of products have been greatly improved, which gives rise to a large number of modern aquatic products processing enterprises specializing in minced fillet products processing, seaweed processing, roasted eel processing, canning and soft packing processing, and dried and frozen products processing. The development of the processing industry has been an important driving force for sustainable fisheries development.

With the development of science and technology, not only can raw aquatic products be processed into aquatic food but many aquatic resources and waste from aquatic food processing can also be processed into fish meal, fish oil, seaweed chemical products, marine health food, marine medicine, leather products, cosmetics, and handicrafts. In particular, marine medicine and health food have been developing very quickly. At present, hundreds of physiological activators have been isolated from marine organisms, and they are believed to have opened a new era for the health and medical treatment of human beings.

The preservation and processing of aquatic products is an important and indispensable component of aquaculture development. Through deep processing and comprehensive multilevel utilization of aquatic products, the value of aquatic resources can be fully utilized, enabling aquatic products to play a greater role in the food supply and human health maintenance.

4.4.2 Status and Development Trends

The traditional technology of aquatic products preservation and processing has been developing for thousands of years; yet, only in the early twentieth century were modern science and technology applied to aquaculture. China has much experience in keeping products fresh using natural ice or other traditional aquatic products preservation methods, such as curing, dry-curing, and fermenting. The main processed products in China are shark fins, trepang, scallops, dried shrimp, salted dried fish, jellyfish, mussels, and dried abalone (Wang 2003). In the *Rites of Zhou,* there were already records of fish dry-curing and sauce making. In *Zhuangzi External Things* and the *Family Sayings of Confucius*, there were also records of "a shop that sells salted fish" and "a market for salted fish." Thus, there are indications that there were dried fish and salted fish products processing and marketing 2000 years ago, or even earlier. Over 2000 years later, drying and salting, the traditional aquatic products preservation and processing technologies, are still in use today. From the early nineteenth century to the 1830s, sealing, heating, sterilization, and preservation technology and artificial refrigeration technology were invented. The 1860s to 1890s period witnessed the development of technologies such as modern refrigeration, cold storage, and ice making, and the canned food-processing industry began to rise. Since the twentieth century, technologies involving refrigeration and cold storage, artificial ice, canned food, fishmeal, and other processing technologies have been introduced to China. However, due to limiting factors of scattered fisheries production areas, strong seasonal fisheries production, and poor industrial technology, by the 1940s, the proportion of refrigeration and cold storage and canned food production in the whole of preservation processing was still far below curing, dry-curing, smoking, and other traditional processing methods (Wang 2003).

The preservation, processing, and comprehensive utilization of aquatic products are still major concerns of fishery countries around the world and a subject of strategic study as well. The sale of fresh and live aquatic products is expected to further increase in the future. Much improvement should be urgently made in the processing and utilization of freshwater fish, low-value fish, and shellfish. General trends of aquatic products processing and utilization are listed as follows (Wang 2003). (1) Quality and freshness should be enhanced (including keeping a small number of aquatic products alive). (2) Diversity, small packaging, and processed products for consumer convenience should be improved. (3) In addition to processing aquatic products into daily food, exploring how to make health food and functional food from them is necessary. (4) Low-value aquatic products and waste should be comprehensively utilized as sources for developing new protein sources and new medicines. Aquatic products (such as marine animals, plants, and microorganisms) with specific pharmacological activities in medicines should be developed, through modern technologies such as separation, purification, structure identification, optimization, and pharmacological effect evaluation. For example, the bioactive peptides derived from marine organisms can regulate organism metabolism.

Future developments of the aquatic products processing industry will not only rely on enhancing and improving traditional technologies, but some advanced modern technologies will also be widely used in aquatic products preservation and processing due to the extraordinary effect of combining modern science and technologies (Chen and Zhou 2018), such as automatic control technology, biotechnology, new heat molding technology, membrane technology, supercritical fluid extraction technology, high pressure technology, high vacuum technology, and aseptic packaging technology. The preservation and processing of aquatic products will play a very important role in the development of the food industry, and it is expected that food varieties will unprecedentedly increase, quality will be greatly improved, and a large number of new products will be developed.

4.5 An Overview of the Science of Fisheries Information Technology

4.5.1 Concept and Research Content

4.5.1.1 Concept

Information technology (hereafter referred to as "IT") refers to technology that can extend the function of the human brain when they are producing, gaining, storing, transmitting, processing, displaying, and using information. With the development of the economy, science, and technology, modern information technology has developed into a highly integrated high-tech field. With modern information science, system science, and cybernetics as the theoretical basis and communication, electronics, computers, automation and photoelectricity as the technological support, IT is a generic term for all modern high technologies, such as producing, storing, transmitting, and processing images, text, sound, and digital information. By the end of the twentieth century, IT was widely used in all activities of national economies around the world and in all areas of society. This technology not only changes the way people work, study, and live but also facilitates profound changes in the industrial structure of human society.

Modern fisheries information technology is a combination of modern information technology and the fisheries industry (Chen and Zhou 2018). By transplanting, digesting, absorbing, transforming, and integrating into the field of fisheries technologies such as computing, information storage and processing, electronics, communication, networks, artificial intelligence, simulation, multimedia, "3S" (GIS-Geographic Information System, RS-Remote Sensing System, and GPS-Global Positioning System), and automatic control, human beings have modernized and informationized the fisheries industry.

Combined with a variety of other new fisheries technologies and integrated with other basic disciplines related to fisheries, such as resources, the environment, and ecology, modern fisheries information technology can express, design, and control fisheries through digitization and visualization of them. Moreover, modern fisheries information technology implements scientific management in the areas of fisheries production, management, circulation, and service and transforms traditional fisheries, thus reasonably utilizing fishery resources, reducing manufacturing costs, and improving the ecological environment (Chen and Zhou 2018). In doing so, modern fisheries information technology can accelerate fisheries development and fisheries industry upgrading, making fisheries development meet the desired goals.

4.5.1.2 Research Content

As we all know, IT consists of three components. First, this field involves information basic technology, that is, manufacturing technology of related materials and components, which is the basis of IT. Second, this discipline involves information system technology, which is the core of IT, that is, technologies relating to obtaining, transmitting, processing and controlling equipment and systems, and information system technology is mainly divided into computer technology, communication technology, and control technology, among others. Third, IT involves information application technology, that ·is, information management, control and decision-making technology, which is the fundamental purpose of IT development. These three components of IT are interrelated and indispensable.

Fisheries information technology is the application of IT in fisheries; thus, it is part of information application technology, and we understand the connotation of fisheries information technology mainly from an application perspective. Fisheries information technology used to refer to the computer technology applied in fisheries. However, with IT development, this term has gradually developed into integrated application involving many new technologies, such as computer networks, microelectronic technology, modern communication technology, databases, computer-aided systems, management information systems, artificial intelligence and expert systems, simulation and virtual reality, multimedia, 3S technology, and automatic control technology.

Thus, fisheries information technology is a technology under constant development, and the connotation of fisheries information technology will be continuously enriched with IT development. Moreover, with the development of society and the profound application of IT in fisheries, the research content of fisheries information technology will be deeply enriched, and its role in promoting fisheries development will also become increasingly significant.

Fisheries information technology is a multidimensional technology system. From the perspective of each industrial structure in the fisheries industry, fisheries information technology involves the information technologies of aquaculture, capture, the processing industry, and fisheries equipment and engineering. From the perspective of fisheries economics and management, fisheries information technology involves the information technologies of fisheries macro-decision making, fisheries production management, fisheries markets, and fisheries science and technology extension and promotion. From the perspective of understanding the development of fishery target laws, fisheries information technology involves the fisheries information technologies of fishery targets and fishery processes. From the perspective of fisheries information itself, fisheries information technology is the integration of obtaining, storing, processing, transmitting, distributing, and expressing fisheries information. From the perspective of the applied form of fisheries information technology, this term is the integration of fisheries management information systems, fishery resources and ecological environmental monitoring information systems, dispatching systems that manage fisheries production and law enforcement processes, fisheries decision-making support systems, fisheries expert systems, precision fisheries systems, fisheries e-commerce systems, and fisheries education and training systems.

4.5.2   Status and Development Trends

4.5.2.1   Status

The history of fisheries information technology development dates back to the days when computers began to be applied in fisheries. In most developed countries, fisheries (especially aquaculture) fall under the category of large agriculture; therefore, to understand the developmental history of fisheries information technology, it is necessary to first understand the developmental history of agricultural information technology.

Agricultural information technology dates back to 1952, when Dr. Fred Waugh, a scientist from the United States Department of Agriculture (USDA), was conducting research on feed mixing. For the next 50 years, there were approximately four stages of development: during the 1950s and 1960s, efforts were mainly for solving scientific computing problems in agriculture, such as the ingredient ratio in feed, statistical analysis of data from field experiments, and planning in the agricultural economy. In the 1970s, due to improved computer storage devices and software technology, various agricultural databases were developed and applied. In the early 1980s, microcomputer technology developed quickly and applying computers in agriculture became a trend, with the focus shifted to research and application of knowledge processing, agricultural decision-making support and expert systems, and automation control. In the 1990s during the Internet network era, virtual agriculture and precision agriculture emerged, based on artificial intelligence, 3S technology, and multimedia technology. Compared with agricultural information technology development, fisheries information technology development in developed countries mainly has three stages:

The first stage: In the 1970s, fisheries information technology mainly served the purpose of scientific computing. For example, the North Pacific Fisheries model, designed by L. J. Bledsoe of the University of Washington, could run on a computer and perform many calculations, greatly improving data-processing speeds.

The second stage: In the 1980s, fisheries information technology mainly served the purpose of data processing and database construction. For example, Basic, Pascal, C, and other programming languages were applied to process data, and the data format was mainly file systems. In the late 1980s, with the emergence of database management systems, such as word processing software packages, Lotus, and so on, a number of fisheries databases were developed and established. For example, marine fisheries biological resource databases, environmental databases, disaster and disease databases, and literature and patent databases were established in the USA, Canada, Japan, and Australia. Since computers were not widely used at this time, those who used computers were mostly software developers, who were not fisheries professionals, resulting in a limited application of computers in fisheries.

The third stage: In the 1990s, with the rapid development of computer technology as representative of IT, fisheries information technology in developed countries rapidly developed, and IT was applied to many aspects, such as government-aided decision making, resource management and environmental protection, water surface utilization and zoning management, weather detection and forecasting, prediction of sea and fishing conditions, fish stock detection, fishing vessel navigation, and real-time command of fisheries production operations. Intelligent expert systems were applied for supervising pond physical and chemical parameters, automatic feeding, feed manufacture, and fish disease diagnosis in aquaculture. The rapid development and wide application of Internet technology promoted the emergence of many fisheries websites and fisheries e-commerce.

At present, all over the world, especially in the USA, Europe, Japan, and other developed countries and regions, fisheries information technology has been widely applied, penetrating into all aspects of fisheries production, management, and scientific research. A brief introduction to fisheries information technology is as follows (Chen and Zhou 2018).

Fisheries Database Systems

A database system (referred to as DS) is a computer system, which can store, manage, and reuse a series of closely linked data sets (databases) in an organized and dynamic fashion. A DS can quantify and normalize information processing by recording, classifying, and sorting a large amount information and storing it in the units recorded in the database. Under unified management of a system, users can query and retrieve data and quickly and accurately access the data needed.

A database is a basis to share fisheries information; therefore, developed countries attach great importance to construction of fisheries databases and development and utilization of information resources. Common fisheries databases are as follows: fisheries resources information databases (germplasm resources and water resources), fisheries environmental information databases (hydrology, meteorology, pests and diseases, and pollution), fisheries production information databases (fingerlings, fish medicines, fertilizers, fishing gear, feed and raw materials), fisheries technology information databases (new technologies, new products, and new species), aquatic product market information databases (sales volume and prices of various aquatic products and global aquatic product markets), fisheries economic databases (fishermen populations, water surfaces, production, fishermen incomes, and employment), fishing vessel and fishing license databases, fisheries policy and regulation databases, and fisheries agency/organization databases.

Many countries have established distinctive, basic fisheries databases. The World Fish Center has established FishBase, the largest fish germplasm resource database in the world, which collects information on 30,000 fish species, covering the majority of information on fish stocks around the world. The FAO and environmental organizations have established global databases of fishery resources, fisheries environments, markets and human resources, such as FiSAT (FAO-ICLARM Stock Assessment Tool), and FISHSTATJ (Software for Fishery and Aquaculture Statistical Time Series). The National Oceanic and Atmospheric Administration (hereafter referred to as NOAA) and National Marine Fisheries Service (hereafter referred to as

NMFS) have compiled the Aquatic Sciences and Fisheries Abstracts (hereafter referred to as ASFA), which provides complete information relating to marine and environmental science engineering, including seven subdatabases, with more than five thousand major journals and patents, conference papers, books, newspapers, and other materials, as well as non-English journals and government reports. Disciplines related to aquatic science are biology, ecology, aquaculture, fisheries, oceanography, limnology, resources and economy, pollution, biotechnology, and marine technology and engineering.

Fisheries Expert Systems

An expert system (hereafter referred to as "ES") is an intelligent computer program, which can reason with acquired knowledge and solve complex problems that only an expert can solve. In other words, an ES is a computer system that has the capacity of to simulate expert decision making and that solves problems by setting knowledge as the center and taking logical reasoning as the approach.

A fisheries ES is based on fisheries professional knowledge, and it is a computer system, which can solve complex practical problems like fisheries experts in a particular field of fisheries. This ES applies the experience of fisheries experts and uses appropriate methods to solve problems with reasoning mechanisms by obtaining, summarizing, understanding, analyzing, and storing knowledge into a knowledge base.

Since the late 1970s, foreign countries have started to apply ES technology to related production areas. At present, this technology is applied to several aspects, such as physical and chemical parameter monitoring in aquaculture waters, automatic feeding, feed manufacturing, fish disease diagnosis, fisheries economic benefit analysis, and aquatic product marketing management. For example, in the early 1980s, a Norwegian company developed fish feeding equipment controlled by a computer system, making bait feeding fully automatic. Japanese researchers, including Kiyoshi Kisui, developed an ES with 28 variables, such as egg abundance, juvenile fish catches and the Kuroshio Current path, and 146 rules on the relationships among these variables, which could predict the status of Japanese anchovy resources in Kanagawa. In 1991, F. Fuchs, a Danish researcher, developed an ES for analyzing the relationship between fish and the environment based on the expert system shell "AUTOKLAS." The FAO developed an interactive ES in 1993, which included an expert knowledge base and a model base. This ES could select and match a surplus production model, which included environmental factors, and it was mainly used to assess and predict fishery resources. In 1995, a research team from the University of Galati in Romania, which consisted of experts in fisheries, computers, and disease pathology, developed the first fish disease diagnosis ES in the world. In 2000, Israel developed a single personal computer (PC) fish disease diagnosis ES, based on fuzzy logic and inference rules. Japanese scholars applied an ES to forecast eel fishing conditions. Karen Hyun and other researchers modeled the long-term fisheries data of 30 Korean fish with an artificial neural network.

Steven Mackinson predicted the distribution structure and dynamics of herring shoals with an adaptive fuzzy ES, which could accurately assess the status of marine fishery resources and played a very important role in rationally and sustainably utilizing marine fishery resources.

Fisheries Management Information Systems

A Management Information System (hereafter referred to as "MIS") is a human-oriented system that gathers, transmits, simulates, processes, retrieves, analyzes, and expresses information through office equipment, such as computer hardware, software, and network communication equipment, to enhance the strategic competition of an enterprise, improve efficiency and benefits, and assist in decision making, control, and management.

A fisheries MIS applies management information system technology to fisheries management, assisting practitioners in conducting fisheries management, making scientific decisions, and improving efficiency. Currently, in developed countries and regions, fisheries MIS has been widely used in fisheries administration management, fisheries production management, fisheries operation management, fisheries enterprise management, fisheries product resources quality management, and fisheries science and technology management.

3S Technology

The term "3S technology" is an abbreviation for RS, GIS, and GPS. 3S combines the technologies of space, remote sensing, and satellite measurement positioning with the technologies of computers, communication, and control. This technology has been widely used in many areas, such as the military, communication, transportation, the environment, territory, and agriculture, playing an extremely important role in sustainable social development.

In fisheries, 3S technology, which emerged in the mid-1980s while developed in the 1990s, was first applied in marine fisheries. Currently, 3S technology has been widely used in predicting fishing conditions, detecting fish stocks, managing fishery resources, monitoring fishing ground dynamics, protecting the environment, real-time predicting and monitoring of various fisheries disasters (e.g., red tides, typhoons, and pests and diseases), utilizing water surface use, zoning management, fishing vessel navigation, and real-time command of offshore production operations. Moreover, with the existing 3S technology, people are researching and developing new 3S technology for different application objects and functions, which plays an important role in fisheries production, scientific research, and management.

Developed Western countries and regions have applied 3S technology in fisheries relatively early. Here are some examples of the early application of 3S technology in the following countries and regions.

In the mid-1980s, some Fisheries Services Center in the USA succeeded in applying remote sensing technology to research on the resource distribution of tuna in the Gulf of California, butterfish and juveniles in the Gulf of Mexico, and research on their fishing grounds. Researchers collected information from the Nimbus 27 CZCS (coastal zone color scanner) watercolor scanner to periodically calculate the spatial distribution of chlorophyll and primary productivity in the Gulf of Mexico. They also calculated the sea surface temperature and graded distribution by combing information from NOAA (National Oceanic and Atmospheric Administration) AVHRR (Advanced Very-High-Resolution Radiometer), and they found correlation between the resource distribution and fishing grounds of butterfish and its juveniles. They developed a quantitative regression model, which, combined with ES, was later widely used in fisheries production in the Gulf of Mexico.

Since the 1980s, the Ministry of Agriculture, Forestry and Fisheries of Japan has gained regular information on fishing grounds and fishing conditions with data from meteorological satellite remote sensing, which has provided the Japanese marine fishing industry with information concerning fishing conditions and sea conditions, on a regular, once every 5/7-day basis (at fixed dates year round), on a regular basis (during the fishing season), or on a once every 10-day basis (year round). Moreover, Tokai University, with the help of satellites, monitored the light distribution of fishing vessels in waters near Japan at nighttime. Combined with an overlay analysis of sea surface temperature of remote sensing inversion, they found that most fishing vessels conducted operation next to the cold-water boundary in the warm–cold water border, which provided the basis for marine fisheries resource management. Currently, varieties of data and ranges of forecasts of sea and fishing conditions are expanding, and Japan takes a lead in 3S technology.

Canada has established the Gulf Geographic Information System (G-GIS) and Marine Environmental Data Systems (MEDS) to help manage domestic and foreign fishing vessels in its 200 nautical mile Exclusive Economic Zone by automatically recording data such as fishing licenses, quotas, catches, and fishing efforts; the UK has developed FishCAM2000, a fisheries production dynamic management system, through combined utilization of GIS, DBMS (database management system), and GPS technologies. FishCAM2000 consists of two parts, a ship-borne system and management system. With the ship-borne system installed in the ship-borne microcomputer, the tailored software system is connected to a global positioning system, and data are gathered in two ways, transmitted automatically or stored in disks. For the management system, the management department can conduct data processing, analysis, and export by connecting ODBMS (Operational Database Management Systems) with GIS. The Skagex electronic atlas, a product of 3S technology jointly developed by Germany, Finland, Norway, Scotland, and Sweden, covers the physical, hydrological, chemical, and biological parameters in the seas of seven Baltic countries.

Computer Networks

A computer network is a system that interconnects multiple computer systems in different locations and with independent functions by using communication devices and transmission media, thus sharing resources and transmitting information in the network through fully functional network software. The Internet is one of the fastest growing, most widely used, and large-scale computer networks. To date, the Internet has covered most of the countries and regions around the world, with tens of millions of networked hosts and hundreds of millions of Internet users. The Internet involves a wide range of information on industrial and agricultural production, science and technology, education, culture and art, business, news, and entertainment. Online shopping, online education, online stock markets, telemedicine, movies on demand, online meetings, and network exhibitions have become a reality. The Internet has become a tremendous asset of human technology and civilization and an inexhaustible global information resource base.

Fisheries information network construction is relatively mature in many developed countries, with numerous quantities and types, wide coverage, distinctive characteristics, versatile information services, strong interactive features, and simple and practical web design. Here are some well-constructed websites with fisheries information networks: the Aquatic Resources website for the Great Lakes region in the USA, Aquatic Resources Information website of Mississippi, and websites run by the FAO Fisheries Department, World Aquaculture Society, Network of Aquaculture Centers in the Asia-Pacific, Aquaculture Information Center of the United States Department of Agriculture, United States Department of Agriculture, and some aquaculture colleges at universities. Traditional fisheries database systems, fisheries expert systems, fisheries management information systems, and other information systems have shifted from stand-alone to networking, and professional fisheries information networks have also been established to provide professional fisheries information services. For example, the United States Fisheries Society, founded in 1870, has established a website, http://www.fisheries.org, providing fisheries information services; aquatic products online market trading, at http://www.fishmark.com, has facilitated fisheries e-commerce, providing online trading opportunities. The United States National Fisheries Information Network has developed the Fisheries Information System (FIS), a nationwide, web-based, unified fisheries information system, providing accurate, effective, timely, and comprehensive fisheries information for the USA and answering questions of who, when, where, what, why and how, which provides decision makers with a basis for fisheries policy and management and decision making, provides researchers with data resources, and provides practitioners with information services.

Multimedia Technology

Multimedia technology refers to the technology that integrates in a single or synthetic form various media information, such as text, audio, video, graphics, images, and animation, with computer processing (e.g., by collecting, compressing, decompressing, editing, and storing). Its essence is not only the integration of information but also the integration of hardware and software. Meanwhile, multimedia technology forms a system with interactive capabilities through logical links. Information processed by multimedia technology is characterized by two important features. One feature is the diversity of information, and the other is interaction in that people can use the keyboard, mouse, touch screen, and other input devices to control a multimedia player through computer software, providing a means that controls and uses information more effectively.

Multimedia technology enriches the means of fisheries information technology, making fisheries information more diversified. Integrated comprehensively with other fisheries information technologies, multimedia technology could enhance the development of multimedia fisheries database systems, fisheries expert systems, and fisheries management information systems. With pictures, text, audio, and images combined, the information would be more accessible to fisheries practitioners. As early as the early 1990s, developed countries had developed and applied multimedia aquaculture management systems, feed formulation expert systems, and fisheries information consulting systems.

4.5.2.2 Development Trends

Fisheries information technology is inseparable from IT. According to the IT development in recent years, the development trends of fisheries information technology are listed as follows (Chen and Zhou 2018).

Networking

Today, networks, especially the Internet, have become a major platform for exchanging and transmitting global fisheries information. Features of resource sharing, high-speed communication, wide ranges, and interaction drive upgrading of network applications. Networks are no longer only for email exchanges; instead, networks facilitate fisheries e-commerce. Moreover, networks are no longer only for information searches; instead, networks serve as a public information service platform like expert systems. The Internet is interwoven throughout almost all aspects of fisheries. For some technical aquaculture problems, fishermen can find solutions from this network and gain relevant technical guidance.

Traditional and single-function fisheries database systems, expert systems, management information systems, and geographic information systems are also gradually shifting to a network environment, thus bringing about greater social and economic benefits. The network environment is also being transformed from LAN (local area network) to WAN (wide area network) IP (Internet Protocol) and from wired to wireless.

Multimedia

The development of high-speed and high-capacity storage technology has further promoted the development and application of multimedia technology, providing various media forms, such as pictures, text, audio, and images for disseminating fisheries information.

In recent years, key technologies have become increasingly practical by applying multimedia network transmission, multimedia databases, multimedia data retrieval, multimedia monitoring technology, and multimedia simulation and virtual reality, and multimedia technology has been widely used in fisheries information by providing multimedia fisheries electronic publications, multimedia expert systems, and multimedia fisheries information consultation systems. In addition, it is widely accepted to disseminate fisheries technology, conduct distance education, and promote technology with multimedia.

Intelligence

Intelligence on fisheries information technology can be detected in two aspects. First, developing and applying various fisheries expert systems are good indications of the wide application of intelligence on fisheries information technology. Second, intelligence technology is increasingly integrated into a wide range of other high technologies. For example, the Chinese Shrimp Farming Expert Decision-Making Consulting System, an expert system developed by the Tianjin Fisheries Research Institute, with support from a basic database and knowledge base, can provide more than thirty technical management decision services, such as shrimp nursery construction, shrimp farming technology, aquaculture breeding decisions, shrimp supporting feed technology, and disease diagnosis and treatment, covering major problems, which occur in shrimp aquaculture production and management, applying pollution-free production technology throughout the aquaculture process.

Integration

With increasing application of databases, management information systems, expert systems, computer networks, multimedia technology, microelectronics, remote sensing, global positioning systems, and geographic information system technology to fisheries, a number of information technologies have been integrated to meet the needs of high-level applications of modern fisheries, which has become a major development trend. For example, "precision fisheries," the technology currently used in fisheries, is the result of integrating a series of fisheries information technologies such as remote sensing, geographic information systems, global positioning systems, fisheries expert systems, and decision support systems.

Virtualization

Virtual reality technology is a comprehensive integrated technology involving computer graphics, human–computer interaction technology, sensor technology, and artificial intelligence. In fisheries, virtual reality technology is mainly integrated with the development and progress of digital simulation, emulation, and virtualization of fisheries technology.

Virtual fisheries can comprehensively utilize computers, simulation, virtual reality, and multimedia technology for breeding virtual aquatic products, providing guidance for farming, and establishing a virtual fisheries resource environment where research can be conducted to study the living environment of aquatic products.

4.5.3 Information Technology and Modern Fisheries

4.5.3.1 Application Status and Prospect Analysis of Geographic Information Systems in Marine Fisheries

Geographic Information Systems (referred to as "GIS") are a newly emerging science integrated with computer science, space science, information science, mapping and remote sensing science, environmental science, and management science. Since the 1960s, GIS has developed into a multidisciplinary basic platform applied in various fields, serving as the basic means and tool to analyze geographic spatial information. At present, GIS has not only become a relatively mature technological science during development but also plays an increasingly important role in all walks of life.

Development of Fisheries GIS

GIS is defined as a combination of computer programming, data, and design, for collecting, storing, analyzing, and displaying geographic reference information. In the early 1960s, GIS as a professional discipline was first introduced in Canada, marking the beginning of an era to solve the space problem with computers. With half a century of development, GIS has now become the primary approach to solving geographical problems in many fields. GIS was first applied in the fields of land resources development, urban planning evaluation, and environmental monitoring. Since the 1980s, this technology has been applied to inland fisheries management and aquaculture farm selection. In the late 1980s, GIS was gradually applied to marine fisheries. Although this technology has been applied to offshore fisheries since the 1990s, covering three oceans, compared with the application on land, its marine application is still severely limited. With a review of the fisheries literature, the features and drivers of the development of GIS and fisheries GIS at each development stage are listed in Table 4.1.

There are three reasons impeding the rapid development of fisheries GIS. First, high costs hinder the development of fisheries GIS in marine fisheries in that a large amount of money is needed to collect data on the biology, physical chemistry, and bottom shape of aquatic organisms. Second, a water system is more complex and unstable; thus, it requires different types of information presented in three or even four dimensions (3D + time). Therefore, the dynamics of a water system imposes a great challenge on GIS technology. Third, many commercial software programs are tailored for land-based use, and they attach importance to the commercial value with support from advanced statistical software. Therefore, these software programs cannot effectively deal with data from fisheries and the marine environment.

Application of GIS in Marine Fisheries

The current application of GIS technology in marine fisheries both home and abroad can generally be summarized in the following main areas: the collection and analysis of fisheries and sea condition data; relationship between fishery resources and the marine environment; aquaculture farm selection; assessment and analysis of fishery resources; tag and release; and forecasting of marine ecosystems and fishing conditions.

- *Collection and Analysis of Fisheries and Sea Conditions*

Data collection is mainly obtaining data by various methods, including with acoustic surveys and remote sensing (satellite or space shuttle). Recently, satellite images and other remote sensing digital information are increasingly introduced to GIS for research on the biodistribution of marine organisms and the dynamic relationship between marine organisms and the marine environment. For example, digital information from acoustic surveys is introduced to GIS to estimate in situ, three-dimensional biomass and map seabed topography and to make further progress in fish ecology. With analysis of altimetry satellite data, researchers are exploring the application to fishery conditions analysis by considering the research findings on the relationship between sea surface height and fishing grounds, and they are determining the theoretical framework of satellite data application to fishery conditions analysis by analyzing according to the theories of oceanography and fisheries science.

- *Relationship Between Fishery Resources and the Marine Environment*

The marine environment is closely related to marine fishery resources, and a good understanding of the marine environment and resource distribution and quantity is important for fisheries management. The spatial distribution of fishery resources and their relationship with the environment are the most basic and common research topics in marine fisheries GIS science, such as GIS mapping and modeling. The application of GIS makes natural resource management more spatially oriented.

Table 4.1 Development of GIS and Fisheries GIS (Gong et al. 2011)

Development Stage	GIS		Fisheries GIS	
	Features	Drivers for development	Features	Drivers for development
1960s	Pioneering period: Expert interest and guide-oriented government; limited to the government and universities; few national interactions	Academic research; application of new technologies; processing of a large amount of spatial data		
1970s	Consolidated development period: Weak data analysis capability; system application and development limited to a certain institute; an increasing influence from government	Resources and environmental protection; rapid development of computer technology; increasing number of professionals		
1980s	Rapid development period: Increasing number of relevant disciplines; rapidly expanding application; commercialization of application systems	Rapid development of computer technology; increasing needs from industry	Pioneering period: Newly emerging technology with a slow development pace; limited application to inland fisheries management and aquaculture farming selection	Development of satellite remote sensing technology; FAO support for GIS; application of land-based GIS technology
1990s	Improvement period: User-oriented era; GIS as an essential office system in many institutions; further deepening of theory and application	Enhanced social awareness of GIS; a significant increase in demand	Rapid development period: GIS widely used in fisheries (from coastal to offshore)	Development of computer technology and increasingly enhanced survey data of marine living resources and the environment
2000s	Expansion period: Society-oriented era; development of social information technology; formation of the knowledge economy	A variety of spatial information related to human daily life	Consolidated expansion period: Consolidation and expansion to more areas (from offshore to deep sea)	Availability and storage of data; common recognition

Traditional models do not consider regional differences, while GIS, as a spatial analysis tool, can tell the difference among regions, because it attaches importance to the interaction among different regions, rather than the average or smoothing value of different regions. GIS modeling is the application of GIS in the process of spatial data modeling. By integrating different data sources, including maps, DEMs, GPS data, images, and tables, GIS can construct various models, such as binary models, exponential models, regression models, and process models, among which exponential models and regression models are commonly applied in fisheries. It is not difficult to construct an exponential model, which is largely used in analyzing habitat suitability and vulnerability, but this method requires GIS users to study digital scoring and weighting. Regression models can combine all independent variables by map overlay computing in GIS, and these models are commonly used in estimating spatial distribution and quantities of fishery resources.

In addition, as for fisheries resource management, the appropriate design of key fish habitats involves spatial measurements. This method is characterized by integrating biotic and abiotic parameters, and it is applicable to all life stages of fish stocks. The significant spatial variation of key fish habitats makes fisheries development and management an extremely difficult issue; thus, GIS, as an efficient spatial–temporal analysis approach, is attracting increasing attention from policy makers, and research on GIS is attracting increasing attention from scientists as well.

To sum up, GIS has been widely used to conduct research on the relationship between fishery resources and the marine environment. GIS will facilitate our understanding of the relationship between resource distribution and the environment, the evaluation of fish habitats, understanding of dynamic resource distribution, and evaluation and management of marine fish habitats.

- *Selection of Marine Aquaculture Farm Sites*

The selection of farm sites is a key factor in any aquaculture type, because it affects the success and sustainability of aquaculture. Selecting farm sites greatly affects capital expenditures, operating costs, productivity, and mortality. GIS is mainly applied in inshore and offshore marine aquaculture and falls under two categories. (1) GIS can be applied in cage aquaculture. In most cases, GIS is applied in the preselection of farm sites in a relatively large area, and the findings of GIS shed light on the further narrowed-down selection of farm sites facilitated by field investigation. (2) GIS can also be applied to inshore shellfish aquaculture. GIS is much more often applied to shellfish aquaculture than to cage aquaculture. GIS involves collecting data on pollution, disease, habitat evaluation with underwater acoustic remote sensing, resource status, carrying capacity, and seasonal mortality.

Marine aquaculture development and management largely involve geographical or spatial considerations, which used to be considered the spatial limiting factors for marine aquaculture. GIS, remote sensing, and mapping play important roles in solving geographic and spatial problems in aquaculture development and management. With satellites and sensors installed in the air, land, and under water, GIS can obtain extensive inshore and offshore data, particularly data on temperature, flow velocity, wave height, chlorophyll concentration, and land and water use.

Essentially, GIS is used to evaluate the feasibility of aquaculture development and establish a framework for aquaculture management. Data problems are the main obstruction in the application of GIS in marine aquaculture, such as the availability of spatial data and attribute data. As for spatial data, there are still many gaps, and they are (1) geographical coverage rate and time gaps; (2) resolution gaps; and (3) data gaps. Most of the research time in marine aquaculture GIS is devoted to determining, collecting, collating, and compiling attribute data and to specifying aquaculture species requirements for the environment and the optimum and restricting factors for aquaculture structures.

- *Fisheries Resource Analysis and Management*

GIS in fisheries resource assessment and management is mainly applied to areas such as marine protected areas (MPAs), reefs, ecosystems, and assessment and prediction. Marine scientists learn about the distribution and abundance of resources through habitat assessment. However, due to the constant spatial and temporal changes in natural habitats, it is difficult to obtain extensive data with traditional methods for data assimilation analysis. GIS can provide an effective tool for solving the internal analysis problem of spatial data, and it can effectively collect, store, display, analyze, and model spatial and temporal data. In addition, GIS can enhance decision making through a combination of different types of data, such as social and political boundaries, sediment type, fish distribution, and so on. GIS can also serve as a supporting tool for MPA decision making. To manage complex MPA issues, MPA managers often seek technology with which a good understanding and analysis of MPA resources and environments is feasible. Currently, GIS and remote sensing are more and more frequently used by MPA managers and researchers to map and analyze resource status and the environment. Many of the world's commercial fish species are facing resources recession and economic decline. The primary recommended remedy is the closure of many marine fishing grounds, and the establishment of MPAs, NTZs (No-Take Zones), and marine reserves are the typical means. Thus, GIS is becoming an integral part in global natural resource management activities.

GIS is a very promising technology for fisheries science and management. Fishery resource analysis and evaluation will enhance fisheries decision making. GIS has a great potential in deciding policies and resource allocation. The former has some influence on decision makers, and the latter has a direct impact on resource utilization. GIS, as a process-modeling tool for policy decisions, has potential (which, currently, has not been recognized), because it is capable of simulating the spatial impact of expected decision-making behavior. Simulation models, especially those dealing with socioeconomic problems, are still in infancy. However, in the future, GIS is expected to play an increasingly important role in this area. Moreover, GIS is also widely applied in decision making for resources allocation.

- *Tag and Release*

GIS has also been applied to research on marine fish tag and release, mainly focusing on multiscale, spatial–temporal research on the relationship between marine organisms and environmental information to grasp the life stage character-istics of marine organisms and their migration paths. GIS is used for multiscale, 3D space-time analysis, but there are few reports or papers on fish dynamic migration using GIS. However, it is a high-profile research method to record fish migration paths and living environments with electronic tags and to conduct spatial–temporal research by combining GPS, GIS, and remote sensing. Application of GIS and remote sensing technology in sea turtle research in North American waters has made revolutionary progress. This research falls under three categories: (1) to track long-distance movements of the turtles using GIS and remote sensing and then determine the turtle population dynamics; (2) to track short-distance move-ments of the turtles and analyze the status of the main habitats and assess reasons for turtle mortality; and (3) to analyze the status of turtle habitats and then to implement sound conservation measures. Multiscale, spatial–temporal analysis is one of the major challenges for marine fisheries GIS, which needs further study.

- *Marine Ecosystems*

As for fisheries science and management, the emergence of the geography of ecosystems and communities indicates the increasing importance of combining different data sources. The research focusing on local areas provides input for multiobjective and multicriteria decision making, and this research calls on GIS as backup. However, GIS is contributive to research only under the following circum-stances when geographical data and environmental knowledge are collected and integrated with standard data. Some researchers set ArcGIS as a platform where marine geospatial ecology tools (MGETs) are integrated. Although the application of GPS, remote sensing, and computers in marine ecology modeling is recognized, researchers insist that ecologists are proficient in using some specific software packages to keep pace with the constant development in the application of GPS, remote sensing, and computers. For example, ecologists should be proficient in using ArcGIS to display and manipulate geospatial data, using R for statistical analysis and using MATLAB to process a matrix, but it is demanding to run these programs independently; thus, they could turn to MGET, an integrated framework of ecologo-geographic processing, where Python, R, MATLAB, and C++ are inte-grated in ArcGIS.

- *Fisheries Forecasting*

Over the past decade, with the improvement of satellite remote sensing informa-tion acquisition, visualization analysis, and mapping technology, the technology of marine fisheries forecasting has rapidly developed, especially the technical means and methods for forecasting the spatial–temporal distribution of some fish species or some fish families, which can then be applied to forecast fishing conditions. The main methods of fisheries forecasting are statistical analysis forecasting (such as linear regression analysis, correlation analysis, discrimination analysis, and cluster analysis), spatial statistical analysis and spatial modeling (such as spatial correlation expression and the spatial information analysis model), artificial intelligence (such as expert systems and artificial neural networks), fuzziness and uncertainty analysis (such as Bayesian statistical theory), and numerical calculation and simulation (such as Monte-Carlo simulation). With an autonomous database, GIS can facilitate the integrated management of spatial–temporal data, spatial overlay and buffer analysis, isoline analysis, exploratory analysis of spatial data, visual display of model analysis results, and vector output of maps. By combining various statistical methods and data on fishing and sea conditions, GIS can facilitate intelligent forecasting of fishing conditions.

4.5.3.2 Marine Remote Sensing and Marine Fisheries

Marine fisheries remote sensing is the application of remote sensing technology in marine fisheries. In marine fisheries, low-altitude aircraft can be used to directly observe and forecast for marine fisheries. Some fish stocks form a certain watercolor and image, some phytoplankton glow under disturbance by fish stocks, and some fish stocks may be under some floating objects, and these are the reasons why the distribution of fish stocks can be forecasted directly from visual observation or photographs taken from low-altitude aircraft. However, there are other means to forecast fishing conditions. Many factors in the marine environment are closely related to fish behavior, such as the water temperature, current, light, salinity, dissolved oxygen, bait organisms, topography, sediment, and meteorology. Electromagnetic waves in various bands from the sea surface reflection, scattering, or spontaneous radiation carry information on sea surface temperature, sea level height, sea surface roughness, and the concentration of various substances in seawater. Since a sensor can measure the energy of electromagnetic waves in different wave bands from the reflection, scattering, or spontaneous radiation of the sea surface, through analyzing electromagnetic wave energy that carries information, researchers can directly or indirectly retrieve some marine physical parameters, such as seawater temperature, chlorophyll solubility, and sea surface height. Through analysis of these marine physical parameters as well as understanding their relationships with fish behavior and fishery resources, researchers can evaluate the status of marine fishery resources and forecast the changes in marine fishing grounds by looking into these marine environmental parameters, thus rationally developing and utilizing marine fishery resources.

Satellite remote sensing technology can continuously, rapidly, and synchronously collect large-scale information on sea surface organisms (e.g., chlorophyll, fluorescence, and primary productivity) and physical parameters (e.g., current, eddy, water temperature, wind, wave, sea surface height, and transparency). The information can be used to assess marine ecological resources and the environment, and high-resolution satellite data can be used to monitor vessels operating in various areas for actual fishing efforts, enhancing the rational development and management of fishery resources. Remote sensing technology can acquire large-scale marine environmental data quickly and dynamically; thus, it has become an important technical method for studying the ocean, and its application in fishing conditions analysis, fisheries management, fisheries resource assessment, and fisheries operation safety has also rapidly developed. Thanks to improved sensor detection ability, the application of remote sensing data in marine fisheries has changed from an initial single-factor application, which is characterized by water temperature data, to a multienvironmental parameter application. Due to the powerful ability of spatial data visualization and spatial analysis, GIS has the ability of spatial positioning; thus, remote sensing data and marine fisheries survey data are integrated on the GIS platform. The integration of 3S will provide a powerful technical platform for marine fisheries research, promoting the development of digital fisheries information. At the same time, the combination of GIS technology, expert systems, and artificial intelligence technology will promote the development of marine fisheries analysis and research toward an intelligent direction.

4.5.3.3 VMS and Fisheries

Overview

The Vessel Monitoring System (VMS) is a fisheries surveillance program proposed by the USA in 1991, in which many countries in the world participate. The equipment installed on fishing vessels can provide information on the location and activities of fishing vessels. Generally speaking, the components of VMS include ship terminals (such as CLS America, vTrack VMS, and BeiDou terminals) for collecting information, communication systems (such as General Packet Radio Service GPRS, Inmarsat, and BeiDou) for transmitting data, and fisheries-monitoring centers for processing and managing such information. A VMS system can be used for different purposes, such as fishing control, scientific research, navigation safety, and marine law enforcement.

Application of VMS Data

The information from a VMS system includes information on the ship identification code, time, longitude and latitude, speed and course, catches, and environment, among which the position of a fishing vessel is the most important. At present, the application of VMS data mainly focuses on three aspects. First, VMS data are used to determine the navigation status of a fishing vessel and to estimate fishing efforts as well. Second, according to a comprehensive analysis of VMS and fishing log data, the statistics and analysis of fishermen's behavior at sea can be conducted, and the scope of fishing grounds can be accurately defined. Third, VMS data can be applied to research on fishing targets and other organisms in related areas, such as the ecological and biological research on whales and seabirds.

Follow-Up Research and Work

A VMS can automatically transmit data on the position, speed, and navigation route of a fishing vessel to a land-based monitoring center, and the fishing vessel can report fishing updates to the monitoring center through VMS, so that the monitoring center can keep informed and monitor the fishing vessel's operation dynamics in real time, which plays a very positive role in monitoring the position, direction, speed, catches, and illegal operations of fishing vessels and detecting possible errors in fishing logs.

In China, the main source of VMS data is the BeiDou satellite system. Compared with AIS (Automatic identification System) and Inmarsat-C systems widely used in other countries, the BeiDou satellite system has the characteristics of fast positioning, high precision, and high frequency of data return. Therefore, with data from the BeiDou satellite system, researchers do not need to bother improving inaccurate data with complex algorithms; instead, researchers can access the navigation route and status of a fishing vessel with a simple method, which helps reduce the consumption of computing resources and improve data accuracy.

The exploration and application of VMS data is still a relatively new subject. At present, research on fishing data is also relatively limited. Most of this research focuses on trawl fishing with relatively simple data processing, and there are few studies on long-line fishing, surrounding nets, and other fishing methods. Therefore, designing more appropriate and accurate methods and models for VMS data processing is still a focus of future research.

4.5.3.4 Big Data and Modern Fisheries

With the rapid development of technologies such as sensors, the Internet, and cloud computing, the amount of data generated by human society is growing explosively, and the era of "big data" has already arrived. China has a large amount of fisheries data, which could provide a solid and reliable basis for government decision making, enterprise management, and scientific research if properly collected, cleaned, integrated, analyzed, and transformed into useful information.

The development of fisheries informationization and big data technology has created fisheries big data. With fisheries big data technology, based on the concept of big data and related technical framework, combined with mathematical models, a large amount of data generated by fisheries informationization can be processed and analyzed and then provided to target users in an intuitive form to solve fisheries problems.

The mass of data generated by fisheries informationization includes information generated at different stages of aquaculture, fishing, processing, supply and marketing, scientific research, management and information on meteorology, water quality, markets, and policy, parameters that have impacts on different fishery stages.

Fisheries data processing and analysis involves data collection, classification, processing, management, exploration, and analysis. Finally, valuable information is extracted and displayed for target users. In short, as for big fisheries data, data are the foundation, analysis is the core, and improving fisheries productivity with information technology is the goal.

Therefore, big fisheries data are an aggregation of all data generated at the different stages of fisheries, such as fisheries planning, production, marketing, management, and scientific research (including all factors affecting these stages, including geography, meteorology, hydrology, environmental protection, policy, and markets). Big fisheries data is also a general term for the technology and application of acquiring, classifying, storing, managing, exploring, and providing fast and valuable services for these data. At present, the mature and widely used big data management and application technology, application structure, data exploration technology, and business management model can be used for reference on the platform of big fisheries data, thus providing effective information support for fisheries planning, production, sales, management, scientific research, and other related services and significantly improving the comprehensive productivity of fisheries.

Big data, cloud computing, the Internet of Things, and other information technologies have been widely used in tertiary industries, such as telecommunications, finance, transportation, and the Internet. These technologies have changed people's way of life. As for primary industries such as agriculture, forestry, animal husbandry, and fisheries, they work with the materials of nature. Therefore, the technology of perception and data collection needs to be improved, and many physical and chemical parameters need to be collected manually, but these needs do not affect the construction of a big fisheries data platform. Data and technological thinking will be the new means of production, tools, and producers. Supported by data analysis tools, fisheries will enter the era of intelligent decision making.

4.5.3.5 Internet of Things and Fisheries

The application of the Internet of Things technology in aquaculture enhances the "intelligent perception and intelligent management" in aquaculture supervision, effectively improving the level of safety supervision. The Internet of Things and big fisheries data integrate sensing technology, wireless communication technology, intelligent information processing, and decision-making technology into all stages of aquaculture, enhancing the intelligent monitoring of aquaculture environments and facilities, automatic control of aquaculture processes, and intelligent decision making on aquaculture production, thus ensuring the safety and reliability of fisheries production.

The real-time measurement and control system of the Internet of Things is used in modern fisheries to realize real-time measurement and control of water quality, which reduces the risks in aquaculture by improving water quality in advance according to the results of water quality monitoring. The application of big data and intelligent decision-making technology facilitates precision feeding and scientific medication to reduce pollution in aquaculture waters and shift the fisheries production pattern from experience-oriented to scientific decision making-oriented. The application of automatic control technology and equipment significantly reduces the labor intensity of fisheries production. Automatic treatment of circulating water improves the utilization efficiency of aquaculture wastewater; greatly improves the yield, quality, and efficiency; and raises the income of aquaculture practitioners.

Modern fisheries Internet of Things provides advanced and practical solutions and technical means for fisheries development in China. This technology facilitates the portability and digitalization of aquaculture information collection, as well as the automation and precision of aquaculture operation. This technology is of great significance for ensuring the high yield, high efficiency, high quality, safety and ecology of fisheries production, changing the production status of fisheries and the aquaculture industry, promoting adjustment of the fisheries industry structure, and promoting the transformation and upgrading of the fisheries industry.

The rapid development of new material technology, microelectronics technology, and micromechanical processing technology is expected to greatly reduce the cost of applying fisheries perception technology and promote the wide dissemination of fisheries perception technology. With the development of new material technology, microelectronics technology, micromechanical processing technology, and optical technology, aquatic information sensing technology no longer only involves laboratory-based physical and chemical analysis but instead facilitates real-time online access to information generated at different stages of fisheries production.

The development of wireless sensor network technology and mobile communication technology has effectively guaranteed reliable transmission of aquatic information. The Internet based on TCP/IP protocol and mobile communication networks based on mobile communication protocol are widely used in data acquisition, long-distance data transmission and control in the modern aquaculture industry, serving as the main channel of data transmission for the fisheries Internet of Things, highlighted by the characteristics of being cheap, stable, high speed, and effective. Mobile communication and the Internet have been exponentially integrated and widely used, playing increasingly important roles in promoting adjustment of the traditional aquaculture economic structure and changing aquaculture modes.

Fisheries big data, cloud computing, mobile communication, and other technologies have effectively improved the level of fisheries information services. With popularization of the application of fisheries big data, the storage, search, data analysis, and calculation of massive data are the key technologies to be solved urgently in fisheries information processing technology. Research on applying fisheries big data will greatly promote the application of information processing technologies such as pattern recognition, intelligent reasoning, complex computing, and machine vision in the fields of precision feeding, intelligent control of aquaculture facilities, disease prediction and early warning, management decision making, quality, and safety traceability, improving the information service level.

4.6   An Overview of Fisheries Economics

4.6.1   Concept and Research Content

4.6.1.1   Concept

Fisheries economics focuses on fisheries production activities, and it is the applied economics studying fisheries production relationships and its development laws. Fisheries industry activities can be classified into four categories: activities related to the ecosystem, fisheries technology systems, fisheries economic systems, and fisheries social systems. Fisheries economic systems and technology systems reflect the nature of fisheries production. Fisheries economic systems and social systems reflect the economic and social nature of fisheries production. Fisheries economics studies the application of general economic laws in fisheries production and the specific economic laws adhering to fisheries production (Chen and Zhou 2018).

The features of fisheries economic activities are mainly listed as follows (Chen and Zhou 2018). First, water resources used in aquaculture are natural resources. Aquaculture has the nature of agricultural production, but the objects live in water; thus, aquaculture is different from the livestock industry and animal husbandry. Second, fisheries economic activities involve various industries. Third, marine fishing has the same nature as agriculture and industry, and most small-scale fishing practitioners are farmers and fishermen along a coast, while large-scale offshore and deep-sea fishing practitioners are generally industrial employees (they are still defined as fishermen), and offshore fishing investment is very extensive. Fourth, the work place for marine fishermen is not fixed but mobile; equipment and fishermen are constantly moving with the flow of fishery resources. Fifth, fishery products must be kept fresh, and due to their perishability, professional collaboration at the stages of production, transportation, processing, preservation, and sales should be coordinated.

4.6.1.2   Research Content

Fisheries economics falls under the category of applied economics. This field focuses on the following research topics (Chen and Zhou 2018): (1) basic theories and methods for research on fisheries economics; (2) macrofisheries economic issues, such as aquatic products trade, fisheries finance, and the development of fishing villages; and (3) microfisheries economic issues, such as aquaculture economics and other fisheries-related cross disciplines, including fisheries technology economics, fisheries institutional economics, and fisheries ecological economics. In a broad sense, fisheries economics is an interdisciplinary cluster involving theories and methods for aquatic science, economics, institutional economics, resource and environmental economics, mathematical economics, econometrics, management, sociology, and politics.

Research on fisheries economics is developing quickly and has the following characteristics (Chen and Zhou 2018). (1) There are currently more interactions among various cross disciplines contributing to disciplinary division and specialization. The basic characteristic of fisheries production is its close interaction and integration with economic reproduction and fisheries natural resources reproduction. Thus, fisheries economics, which focuses on "issues of fisheries, fishing villages, fishermen and laws in fisheries economic activities," involves the cross disciplines of aquatic science, economics, resource economics, and management, and their interaction and integration are further enhanced. (2) Research on fisheries economics is increasingly multilevel and multiangle. There are more interactions and integration between micro- and macroanalyses, normative and empirical analyses, and qualitative and quantitative analyses, and there are more updated research methods. (3) The research scope of fisheries economics is continuously expanding, and new theories and practices call for innovation in fisheries economics research and more comprehensive and systematic fisheries research. All these factors are challenges to the design of fisheries economics disciplines, the relevant education, and research.

Fisheries economics has formed a discipline system and an integrated research team. However, as some studies have indicated, there are too few full-time researchers in fisheries economics, and research efforts are still weak and unevenly allocated. To match with a fast-growing fisheries economy, more systematic and consistent research is needed, and more research is to be conducted on applying the theories and methods of fisheries economics.

4.6.2 Overview and Development Trends of Fisheries Economics

Fisheries economics is a science formed with the development of a capitalistic commodity economy in the fisheries industry. In 1776, Adam Smith, a British classic economist, conducted a comprehensive analysis on the geographical conditions of oceans, rivers and lakes, fisheries investment, and the impact of fishery costs on aquatic products pricing. In his masterpiece of *An Inquiry into the Nature and Causes of the Wealth of Nations*, Smith illustrated the fisheries risk with a review of an anecdote: in 1724, a British fisheries company started off whaling even though only one out of eight attempts of whaling was profitable. Smith stated in his masterpiece that whoever were involved in fisheries should be ready to consider risks and profits. In the middle of the nineteenth century, elaborations were made on fisheries economic activities in Marxism economics. Marx spoke highly of the role of aquatic products in activating human brains, and he classified the fishing industry as an extractive industry, viewing fishing operations as labor that created surplus value. What Marx argued about fisheries contributed to the development of fisheries economics. In the early twentieth century, systematic studies were carried out on fisheries economics, and, *Fisheries Economics*, the fisheries economics monograph in the early days, was written by a Japanese scholar, NinagawaTorazo, in 1933. In 1961, Okamura Kiyozo, another Japanese scholar, re-edited *Fisheries Economics*, which was then republished many times after 1972. There were also other important books on fisheries economics, and these included *Fisheries Economics Theory*, written in 1979 by a Japanese scholar, Kondo Yasuo, and *Fisheries Economics Theory*, coauthored in 1982 by TeruakiKumitsu and his colleagues. *Fisheries Economics Theory* conducted an economics analysis on the balance of the means of production, market mechanisms, income distribution, and fisheries economic structure based on a detailed elaboration of the process of fisheries production, aquatic products circulation, and consumption (Chen and Zhou 2018).

In countries with developed fisheries economies, such as the USA, Canada, and Russia, researchers have conducted studies into the fisheries economy on the precondition that the fisheries economy is sectoral economics in either an extractive industry or industry. For example, Jean Sawyer, a fisheries economist from the Soviet Union, wrote the *Soviet Union fisheries economics*, in which he studied the fisheries economy, viewed it as a component of the industrial economy, and discussed the status and role of fisheries in the national economy, performed an economic evaluation of aquatic resources and improvements of fisheries science and technology, and presented theories and methods for fisheries economic management. Nordic countries, such as Norway, are focusing on in-depth studies on the fisheries economy from the perspective of fisheries resource economics. Then, they evaluate the economic effects of fishing and propose significant technical and economic measures to impose fishing restrictions.

In China, where the aquaculture industry is more developed, fishing is treated as an important component of large-scale agriculture; thus, fisheries economic management has been seen as a component of agricultural economic management. In China, fisheries economics is an applied economic science and develops based on the experience of the fishing industry, both at home and abroad. The reform and opening-up has promoted the in-depth study of fisheries in many areas, and remarkable achievements have been made. More in-depth studies have been performed in the following aspects: economic analysis of fisheries development, development of the fisheries industry and its restructuring, aquatic products trade, industry associations and cooperative organizations, fisheries insurance, fisheries modernization, sustainable utilization of fishery resources, and fisheries development strategies.

Prospective research on fisheries economics will highlight the following topics (Chen and Zhou 2018). (1) Fisheries development strategy and planning, such as the development strategy of modern fisheries, development strategy of fisheries under ecological civilization, and development strategy of marine fisheries. (2) Steady fisheries development and its technological innovation. Fisheries development will encounter many problems, such as the exhaustion of fishery resources, aquatic environmental and aquatic product safety, rising labor costs, and changes in international trade patterns, but technological innovation plays a key role in solving these problems. Therefore, restrictions on output growth and technological innovation alternatives should be the priority topics of future research. (3) Aquatic products trade and subsidy policy. With the universal economic depression all over the world, fisheries trade barriers and trade conflicts may be aggravated, the international competitiveness of aquatic products should be strengthened and further studies should be conducted on how to take advantage of international trade rules to support aquaculture development strategy. (4) Resource-saving, environment-friendly, and healthy aquaculture with safe products. Further research should be conducted on sustainable aquaculture development models. (5) Control over fishing efforts and the livelihoods of inshore fishermen. Protection of the rights and interests of vulnerable fishermen, fisheries development, and traditional fishing zones are listed under this topic. (6) Policy-based fisheries insurance model. (7) Public administration, modernization, and information construction of fisheries in China. (8) Sustainable fisheries development models.

References

Cardrin SX, Kerr LA, Marianl S (2014) Stock identification methods. AcademicPress, Cambridge

Chen XJ (2014) Fisheries resources and fisheries oceanography. China Ocean Press, Beijing. (in Chinese)

Chen XJ, Zhou YQ (2018) Introduction to fishery. China Science Press, Beijing. (in Chinese)

FAO (1957) Modern fishing gear of the world I. Fishing News1959

FAO (1963) Modern fishing gear of the world II. Fishing News1965

FAO (1970) Modern fishing gear of the world III. Fishing News1973

FAO (1988) Proceeding of fishing gear and fishing vessels Design of the World. St. Johns, Cananda

GarciaSM (2011)Selective fishing and balanced harvest in relation to fisheries and ecosystem sustainability. Report of a scientific workshop organized by the IUCN-CEM fisheries expert group (FEG) and the European Bureau for Conservation and Development (EBCD) in Nagoya (Japan), 14–16 October 2010.Gland, Switzerland, and Brussels, Belgium, IUCN and EBCD. p33

Gong CX, Chen XJ, Gao F, et al. (2011). Development and application of geographic information system in marine fisheries. Journal of Shanghai Ocean University, 20(6): 902–909 (in Chinese)

Guan WJ, Tian SQ, Zhu JF et al (2013) A review of fisheries stock assessment models. J Fish Sci China 20(5):1112–1120 (in Chinese)

Leah Zelditch M, Swiderski DL, David Sheets H (2012) Geometric Morphometrics for biologists (second edition). AcademicPress, Cambridge

Smith ADM, Brown CJ, Bulman CM et al (2011) Impacts of fishing low-trophic level species on marine ecosystems. Science 333(6046):1147–1150

Wang ZH (2003) Aquatic products industry and utilization. Chemical Industry Press, Beijing. (in Chinese)

ZhouS (2013). Balanced harvest: an innovative solution for biodiversity conservation and sustainable fisheries. Commonwealth Scientific and Industrial Research Organization, Australia

Chapter 5
Sustainable Development and Blue Growth of Fisheries

Xinjun Chen and Yinqi Zhou

Abbreviations

$CaCO_3$	Calcium carbonate
GCMS	General circulation model
GNP	Gross National Product
HCO	Bicarbonate
IUCN	The International Union for the Conservation of Nature and Natural Resources
MEY	Maximum Economic Yield
MSY	Maximum sustainable yield
UNCED	UN Conference on Environment and Development
UNEP	United Nations Environment Program
WCED	The World Commission on Environment and Development
WWF	The World Wildlife Fund

5.1 Overview of Sustainable Development Theory

Since the 1970s, in light of reflection on the population, resource, and environmental problems caused by the traditional economic growth mode, sustainable development has been proposed as a new social and economic development strategy. The transformation of development strategy means a transition of the economic growth mode and resource allocation mechanism.

X. Chen (✉) · Y. Zhou
College of Marine Sciences, Shanghai Ocean University, Lingang New City, Shanghai, China
e-mail: xjchen@shou.edu.cn; yqzhou@shou.edu.cn

© Science Press & Springer Nature Singapore Pte Ltd. 2020
X. Chen, Y. Zhou (eds.), *Brief Introduction to Fisheries*,
https://doi.org/10.1007/978-981-15-3336-5_5

5.1.1 Transformation from Economic Growth to Economic Development and to Sustainable Development

5.1.1.1 Transformation from Economic Growth to Economic Development

Since the industrial revolution, people have been obsessed with the pursuit of economic growth. Economic growth refers to an increase in per capita GDP within a certain period and, sometimes, an increase in personal income or actual consumption. Especially after the Second World War, the economic growth strategy was the dominant social and economic development mode for almost half a century. One feature of the economic growth strategy is that people have constantly increased the breadth and depth of resource development and utilization and have supported the rapid growth of GNP (Gross National Product) or per capita GNP through massive resource consumption with scientific and technological progress. Global economic growth entered its golden era during the 1950s–1960s. For instance, from 1963 to 1968, the average growth rate of the total world's industrial production was 7% per year, with a per capita growth rate of 5%. Indeed, people have benefited from the economic growth strategy: rapid GNP growth has increased the economic level of the whole world, and the economy and overall national strength of each country have been strengthened. Moreover, per capita GNP growth has contributed to the elimination of poverty in many countries, so people's income and consumption levels have dramatically increased and their consumption patterns have developed to a more comfortable and higher level.

However, although the economic growth strategy has constantly improved the economic level and per capita income, it has also caused a series of problems: the inequality between the rich and poor is escalating, which has caused diminishing opportunities for good education and health welfare for a considerable number of people, giving rise to social contradictions including unemployment, deteriorating health, rising crime rates, and so on and leading to the emergence of social instability. Therefore, in the 1970s, the concept of the economic development strategy was proposed to replace the economic growth strategy. Economic development refers to development that enables a series of social goals to be fulfilled. From this point of view, a development cannot be called an economy of development if it fails to change the social and economic structure in spite of its capability to continuously enhance per capita income over time. A series of social goals changes over time, so economic development is said to be a process of social progress. Economic development usually consists of three interactive means of enhancements/improvements.

1. Improvement of personal or social welfare. This improvement includes enhanced per capita income (especially for developing countries) and improved environmental quality, as per capita income and environmental quality jointly determine the level of individual or social welfare.
2. Improvements in education, health, quality of life, and so on. This enhancement represents progress in skills, intelligence, capability, and so on.

3. Enhancement of a country's self-respect and self-love awareness. The social and economic development of a country should also demonstrate an improved sense of independence.

The connotation and extension of economic development are much broader than those of economic growth. Economic development not only contains the capacity for economic growth but also relates to improved social welfare or social progress. In the transformation from economic growth to economic development, people have begun to go beyond the simple pursuit of economic growth, gradually transferring development goals to improvements and changes in social and economic structures.

5.1.1.2 Transformation from Economic Development to Sustainable Development

Compared with economic growth, economic development endeavors to harmonize economic goals with social goals, but it has the same defect as economic growth—ignorance of the role of resources and environment in economic development. This ignorance has led to accelerated resource depletion and environmental deterioration in the process of economic growth, mainly manifested as follows.

1. Natural resources have suffered unprecedented levels of damage and are gradually being exhausted. Global soil has experienced rapid deterioration as well. Over the past 50 years, the productivity of more than 1.2 billion ha of land (larger than the combined size of China and India) has dropped significantly. Meanwhile, nearly six million hectares of production land turns into barren land every year, and more than 11 million ha of forests have been destroyed. Between 1980 and 1990, forests have decreased by 11.6%, giving rise to soil erosion, land desertification, and climate change. The shortage of water resources is believed to be one of the bottlenecks that limit the survival and development of human beings all over the world. Due to the uncontrolled exploitation of nonrenewable resources, mineral resources are being depleted. Since the Second World War, global resources consumption has doubled every 14 years, contributing to the possible complete depletion of oil and natural gas, two highly significant resources for human beings, in half a century. Moreover, excessive resource utilization and environmental deterioration have led to alarming rates of reduction and extinction of biological species—the world's most important resource pool. In 1989, Edward Wilson, a famous biologist, estimated that 50,000 species go extinct per year, 10% of the world's advanced plant species are threatened, and 3/4 of bird species are gradually decreasing in number and, thus, under threat of extinction. He also estimated that 0.5–1 million animal species would disappear from the earth by the end of the twenty-first century. Thus, the loss of species and biodiversity leaves human beings in a very dangerous situation: the biological chain on which humans depend for survival is shrinking and shortening.
2. The environment is suffering from increasingly serious pollution pressures. The massive discharge of various types of waste during economic production and

human life results in deterioration of the global environment, expanding from the initial stage of regional air pollution, water pollution, solid waste, and noise pollution to the second stage of global environmental problems: global warming, ozone depletion, acid rain, and biodiversity decline. Due to excessive fossil fuel consumption, CO_2 emissions have increased dramatically, causing the greenhouse effect and sea level rise. The consequences are very destructive. The early twenty-first century will witness the deterioration of existing agricultural production land, the submergence of coastal cities, and even the collapse of national economic systems in coastal countries. Moreover, the stratospheric ozone depletion caused by excessive discharge of industrial waste gas would dramatically increase the incidence of cancer in both human beings and livestock and even endanger the marine food chain. Furthermore, emissions of sulfur oxides and nitrogen oxides are rising in most areas of the world, and acid rain is spreading across the globe, leading to forest mortality, damage of water systems and lake soil, and the disappearance of national art and architectural heritage.

Overall, it is increasingly clear that the destruction and exhaustion of resources ruins future economic growth, while environmental deterioration hinders economic growth and social welfare enhancement. It is equally clear that the existing economic development mode is eroding the resource and environmental foundation, which is a significant prerequisite for human survival and economic development. People are beginning to worry about the sustainability of current economic growth. Thus, people have to reflect on and summarize the defects and deficiencies of the traditional economic development mode, striving to find a long-term, sustainable, and future-oriented model with heightened awareness of the sustainability of the environment and natural resources. As a result, a new development theory named sustainable development has been proposed since the 1970s.

The emergence and development of sustainable development theory has gone through a step-by-step advancement. From the emergence of the theory to the formation of theoretical systems and, finally, to recognition as a development strategy, this theory has experienced roughly four vitally important milestones.

The first milestone is the declaration of extensive ecological damage and environmental pollution caused by improper resource use at the UN Conference on the Human Environment in 1972 in Stockholm, calling for an awareness that the economy and environmental development must be coordinated. This was the first time that human beings recognized the interrelationship between economic development and resource environmental protection across the world. Although a clear idea of sustainable development was not put forward at the conference, this event made people realize that resources and the environment play a very important role in economic development. The idea of sustainable development emerged from a discussion of the relationship between the environment and development. Therefore, this conference was regarded as the first milestone in the initiation of sustainable development.

Then, the second milestone is the World Conservation Strategy released by the International Union for the Conservation of Nature and Natural Resources

(IUCN) and the World Wildlife Fund (WWF) in 1980. This strategy calls on the world that "it is necessary to harmonize the relationship among nature, society, ecology, and economy in the process of using natural resources to ensure global sustainable development." Thus, this was the first international document to propose sustainable development.

Next, the release of the research report named Our Common Future is regarded as the third milestone of sustainable development. This report was published in 1987 by the World Commission on Environment and Development (WCED), then headed by Gro Harlem Brundtland. In light of an objective analysis on experiences and lessons learned in social and economic development all over the world, this research report clearly defined sustainable development and put forward a global sustainable development strategy and countermeasures for the phase before 2000 as well as for the future in the twenty-first century. In this report, sustainable development is defined as "development that not only meets the needs of the present but also refrains from jeopardizing future generations' ability to meet their needs." Therefore, sustainable development should first meet the basic needs of all people, especially the poor, ensuring everyone's desire for a better life. Then, sustainable development should require technology and social organizations to impose restrictions on the ability of the environment to meet present and future needs. At the very least, economic development should not jeopardize the natural system consisting of the atmosphere, water, soil, organisms, and more that supports life on earth. This definition of sustainable development has obtained extensive approval and acceptance worldwide, sparking a heated global debate on sustainable development problems, while promoting the formation and maturity of the sustainable development theoretical system as well.

Finally, the fourth milestone is the UNCED in 1992 in Rio de Janeiro, Brazil, where programmatic documents were adopted, including the *Rio Declaration*, *Agenda 21*, *Convention on Biological Diversity* and so on, an agreement was reached on the key components of sustainable development, and awareness was raised on the interrelationship between the environment and economic development. The *Agenda 21* adopted at this conference was a broad action plan, providing a blueprint for action and calling on every country to implement the plan through policy making and strategic decisions. As an important milestone, the UN Conference on Environment and Development signified that the proposal on sustainable development was ready for implementation and that sustainable development was recognized as a practical goal commonly pursued by human beings.

Actually, two great leaps in human cognition have been made in the process of transformation from economic growth to sustainable development. The process from economic growth to economic development expands people's economically oriented cognition to society-as-a-whole-oriented cognition with an awareness of the importance of harmony between economic goals and social goals. Thus, the sustainable development proposal has managed to make people recognize the role and status of resources; the environment in social and economic development; the importance of a dynamic balance between resources and the environment; and the necessity of unity among the economy, society, and environment. Thus, recognizing the difference between economic growth and economic development is a breakthrough and development in economics, bringing about great changes to traditional economics.

5.1.2 Concept of Sustainable Development

With the recognition of sustainable development problems, people have had a long and extensive discussion about the basic concepts of sustainable development. Sustainable development involves all aspects of social and economic development; people have different interpretations through different perspectives (Chen 2004).

5.1.2.1 Representative Viewpoints

1. Sustainable development is defined as ecological sustainability with a focus on natural attributes. Ecological sustainability attaches importance to the ecological balance between natural resources and their development and utilization to meet the growing demand for ecological resources brought about by social and economic development. For instance, at the symposium on sustainable development jointly hosted by the International Association for Ecology and International Union of Biological Sciences in November 1991, sustainable development was defined as the "production and renewal capacity to protect and strengthen environmental systems." In addition, based on the concept of the biosphere, some scholars consider that sustainable development seeks an optimal ecosystem to support the ecological integrity and fulfillment of human aspirations to create a sustainable living environment for human beings.
2. Sustainable development is defined with a focus on social attributes. For example, in 1991, Care for the Earth—A Strategy for Sustainable Living, a report jointly released by the International Union for Conservation of Nature, United Nations Environment Program and World Wildlife Fund, defines sustainable development as "improving the human quality of life on the condition of survival within the capacity to sustain an ecosystem." Furthermore, this report highlights that the ultimate goal of sustainable development is to improve the quality of life and create a better living environment for human beings. Thus, this concept of sustainable development highlights that social equity is the mechanism and goal of sustainable development strategy. In consequence, the nature of "development" includes improving human health; enhancing quality of life; obtaining access to required resources; and creating an environment that guarantees equality, freedom, and human rights for people.
3. Sustainable development is defined with a focus on economic attributes. Considering economic development as a foundation of national power and social wealth, sustainable development encourages economic growth instead of restricting

economic growth in the name of environmental protection. However, sustainable economic development emphasizes not only the quantity of economic growth but also the quality of economic growth, so it is important to realize the coordination and unification of economic development and ecological environmental elements. Sustainable economic development should not be achieved at the cost of the ecological environment. For example, some scholars define sustainable development as "the means to maximize the net benefits of economic development on the precondition that the quality of natural resources and the services that they offer are protected." Other scholars maintain that sustainable development means "today's resource usage should not decrease future revenue."

4. Sustainable development is defined with a focus on scientific and technological attributes. In addition to policy and management factors, scientific and technological progress plays an important role in implementing sustainable development. Without the support of science and technology, the sustainable development of mankind would not be feasible. Therefore, some scholars expand the definition of sustainable development from a technological perspective, and they consider that sustainable development means switching to more environmentally friendly and more effective technologies, using "zero emissions" or "enclosed" processes as much as possible and minimizing the consumption of energy and other natural resources as much as possible. Other scholars argue that sustainable development is feasible by creating technical and technological systems in which less waste and pollutants are discharged. These scholars believe that pollution manifests poor technology and low efficiency and that it is not an inevitable result of industrial activities. Moreover, they advocate for technical cooperation between developed and developing countries to narrow the technological gap and increase the economic productivity of developing countries. Meanwhile, they suggest that we should develop technologies by using mineral energy more efficiently; provide safe and economical renewable energy technologies to limit carbon dioxide emissions that warm the global climate; and prevent the production and usage of certain chemicals with proper technology to protect the ozone layer, to gradually solve global environmental problems.

5.1.2.2 Views Widely Accepted by the International Community

The above-mentioned concepts are the popular concepts of sustainable development, but they are defined in such a narrow sense that cannot be universally recognized by the international community. In 1987, in light of a systematic survey and study of the world economy, society, resources, and environment, Brundtland, the former prime minister of Norway as well as the host of the World Commission on Environment and Development (WCED), released a long special report entitled Our Common Future at the Commission meeting. In this report, sustainable development is defined as development that not only meets the needs of the present but also does not jeopardize future generations' ability to meet their needs. The Statement on Sustainable Development was passed at the meeting of the 15th Session of the United

Nations Environment Program (UNEP) Council in May 1989. Sustainable development is defined in this statement as development that not only meets current needs without weakening the ability of future generations to meet their needs but also refrains from interfering with national sovereignty. The UNEP believes that sustainable development requires domestic cooperation and international parity, including assistance to developing countries in accordance with the order of priorities and development goals of these countries' development plans. In addition, sustainable development means a supportive international economic environment, which not only leads to sustainable economic growth and development in every country, especially developing countries, but also has a positive influence on good environmental management. Sustainable development also means maintaining, rationally using, and improving the natural resource base that supports ecological stability and economic growth.

Therefore, Brundtland's definition of sustainable development, "meeting current needs without weakening the ability of future generations to meet their needs," has been the concept of sustainable development universally accepted by the international community. The core of this definition is that healthy economic development should be based on ecological sustainability, social justice, and active participation of people in their own developmental decisions. The goal of sustainable development is satisfying the human need for personal all-round development and protecting resources and the ecological environment without jeopardizing the survival and development of future generations. This goal pays special attention to the ecological rationality of various activities and highlights that we should encourage economic activities that are conducive to resources and the environment, while abandoning the opposite activities.

5.1.2.3 Basic Principles of Sustainable Development

If we implement the new sustainable development model for human beings, that is, forming a sustainable and efficient coordinated operation mechanism in the ecological environment, economic growth, and social development, we must follow the three principles of fairness, sustainability, and commonality.

1. Principle of equality. The principle of equality refers to the equality of opportunity; thus, the equality principle required by sustainable development includes three levels: the first level is fairness for the present generation, that is, horizontal equity among people of the same generation. Sustainable development should meet basic human needs and give the population chances to satisfy their desire for a better life. However, frankly speaking, currently, the reality of society is that some people are rich while others, especially those who make up one-fifth of the world's population, live in poverty. It is impossible for us to achieve sustainable development if the world has such extreme disparity and polarization between the rich and poor. Therefore, for the sake of the world's fair distribution and equal development rights, we should prioritize poverty alleviation in sustainable

development. The second level is intergenerational equity, that is, the vertical equality among people of different generations. We have to recognize that natural resources, which human beings live on, are limited and that the contemporary should not destroy natural resources and the environment—conditions for generations of humans to meet their needs— for the sake of our own development and needs. In short, every generation should be given equal rights to use natural resources. The third level is the equable distribution of limited resources. At present, the distribution of limited natural resources is uneven: developed countries, which account for 26% of the world's population, consume over 80% of the world's energy, steel, and paper, while developing countries face serious resource constraints in economic development.

From this perspective, sustainable development not only needs to achieve equity among contemporary people but also equity between contemporary and future generations to provide all people with an equal opportunity to have a better life. Ethically, future generations should have the same rights to resource access and the environment as contemporary generations. Compared with future generations, the contemporary generation is in an uncompetitive dominant position that is similar to monopolizing resource development and utilization. Therefore, sustainable development requires that the contemporary generation should not only take their own needs and consumption into consideration but also take up the historical morality and responsibility to consider the needs and consumption of future generations. Generational equity means that neither of these generations is in a dominant position; that is, each generation has equal opportunities for development.

2. Principle of sustainability. The core of the sustainability principle is that the economic and social development of human beings cannot exceed the bearing capacity of resources and the environment. Therefore, resources and the environment are the basis and preconditions for human survival and development. Furthermore, the primary preconditions for sustainable development are sustainable resource use and the maintenance of ecosystem sustainability. Meanwhile, sustainable development requires people to tailor their lifestyles based on sustainable conditions and determine standards for individual consumption within ecological resilience. In consequence, the principle of sustainability reflects the principle of equality in sustainable development.

3. Principle of universality. Considering the differences in the history, culture, and development among countries, there will be no universal agreement on a wide variety of specific goals, policies, and implementation steps for sustainable development. However, sustainable development is regarded as the overall objective of global development, so we should jointly comply with the principles of equality and sustainability embodied by sustainable development. In addition, the key to this objective is a unified global effort. In a broad sense, sustainable development strategy promotes harmony among human beings as well as between human beings and nature. If everyone would consider the impacts of their actions on others (including on future generations) and the ecological environment and behave according to the principle of "universality," a mutually

beneficial relationship could be maintained among human beings as well as between human beings and nature; thus, sustainable development can be achieved.

5.1.3 Basic Features of Sustainable Development

Compared with the traditional mentality on development and environmental protection, the latest mentality on sustainable development has distinct features, while understanding these features is of great importance in understanding the content of sustainable development. In general, sustainable development has three basic features.

1. Because economic growth is the foundation for national power and social wealth, sustainable development encourages economic growth. At the same time, sustainable development not only attaches great importance to the quantity of growth but also strives to improve quality, enhance efficiency, save energy, reduce waste, change traditional production and consumption patterns, and implement environmentally friendly production and civilized consumption modes.
2. Sustainable development should be based on nature protection and coordinated with the bearing capacity of resources and the environment. Therefore, in the process of development, we must protect natural resources and the environment through controlling pollution, improving environmental quality, protecting life support systems, maintaining biodiversity, preserving the integrity of the earth's ecology, and ensuring the sustainable use of renewable resources, to keep development within the earth's bearing capacity.
3. Sustainable development should aim to improve the quality of life and adapt to social progress. The fact of modern social economic development is that most of the people in the world are semipoor or poor. Thus, sustainable development must be linked to solving poverty problems for the majority of the population. As for developing countries, poverty and underdevelopment are two significant reasons for resource extinction and environmental collapse. Only by eliminating poverty could we enhance our ability to protect and develop the environment. Because different countries in the world are at different development stages, the specific goals of their development are different as well, but the development connotations should share much in common, that is, enhancing the quality of human life, improving human health, and creating a social environment that guarantees equality, freedom, educational opportunities, human rights, and protection against violence.

The above three features indicate that sustainable development embraces ecological, economic, and social sustainability, which are interrelated and indivisible. Actually, the isolated pursuit of economic sustainability inevitably leads to economic collapse, while the isolated pursuit of ecological sustainability does not ultimately prevent global environmental recession. Above all, taking ecological sustainability as the foundation, taking economic sustainability as the condition, and taking social sustainability as the goal, human beings should keep the sustainable, stable, and healthy development of the complex natural–economic–social system as the ultimate goal.

5.2 Basic Theories of Sustainable Utilization of Fishery Resources

5.2.1 Characteristics of Fishery Resources

Fishery resources are also known as "aquatic resources," known collectively as species and quantities of economic animals and plants with development and utilization value in natural waters, including fish, crustaceans, mollusks, mammals, algae, and so on. All of the above resources are important sources of human food and the material basis for aquatic industry development. In addition, the situation of fishery resources varies with the biological characteristics, habitat environment, and human utilization. Fishery resources have the following main characteristics (Chen 2014; Chen and Zhou 2018).

5.2.1.1 Regeneration

Fishery resources are biological resources that can proliferate by themselves. The individuals or populations of organisms keep resources constantly recruited through reproducing, developing, growing, and replacing the old with the new. Meanwhile, populations of organisms stay relatively stable through a certain self-regulation capacity. Moreover, fish stocks can also be maintained or restored by artificial farming, proliferation, and artificial release. However, overfishing and deterioration of the ecological environment contribute to the damage of fishery resources. The insufficient or decreasing number of recruitment groups eventually leads to the recession or even extinction of fishery resources.

5.2.1.2 Mobility

Most aquatic animals have migratory habits for feeding, reproducing, overwintering, and so on, including anadromous salmon, catadromous eel, oceanic migratory tuna, seasonally migrating *Larimichthys crocea, L.polyactis*, hairtail, and so on. Many species migrate and inhabit multiple regions or waters under the jurisdiction of different states. Therefore, fish migration makes it difficult to specify resource ownership. As a matter of fact, the situation of "whoever catches owns" occurs, that is, "possession is owned" applies to public resources, which means the sharing of fishery resources, that is, the externalities of economics. These characteristics bring about particularities and difficulties in fisheries management, including plunder and waste of fishery resources in exploitation and utilization, as well as excessive investment in capacity development for priority possession. In addition to fish, migratory and flowing resources also include human beings, birds, insects, air, water, oil, and so on. Furthermore, the migratory and flowing characteristics of these resources have common ground in management; in other words, we can use the above management methods and experiences to manage fishery resources.

5.2.1.3   Volatility

Fishery resources refer to biological resources living in waters directly affected by the aquatic environment. Therefore, cyclical changes in the earth's climate and the marine environment may cause fluctuations in fishery resources. In the process of biological reproduction and evolution, the instability and interaction of different ecosystem components also cause quantitative fluctuations. Human activities including fishing also have a significant impact on the declining number and structural changes of fishery resources. Therefore, the reasonable development and utilization of fishery resources is an important key to the sustainable development of the ecological environment.

5.2.1.4   Elusiveness

Fishery resources, such as fish, shrimp, shellfish, and algae, inhabit waters where they are distributed in dense grasses, lakes, or oceans. Furthermore, they move around occasionally, so it is difficult to locate them and conduct surveys for the sake of statistics. The elusiveness of fishery resources makes it difficult to assess fishery resources and explore fishing grounds, which causes great uncertainty in the knowledge of population numbers and habitat situations.

5.2.1.5   Variety of Species

There are many varieties of fishery resource species, including fish, crustaceans, mollusks, marine mammals, algae, and so on. (1) Fish are the largest group of fishery resources. There are approximately 21,700 fish species in the world, only approximately 100 of which are major fishing targets. (2) Crustaceans mainly include shrimp and crabs. There are more than 3000 shrimp species, which mainly inhabit the ocean. (3) There are approximately 100,000 mollusk species, half of which inhabit the ocean, comprising the largest family of marine animals. The cephalopods squid and cuttlefish and the bivalves oysters and mussels are some of the popular mollusk species. (4) Marine mammals include cetaceans, seals, sea otters, dugongs, manatees, and so on, most of which fall under the protected species category. (5) There are 2100 genera and 27,000 species of algae, which are widely distributed not only in rivers, lakes, and seas but also in moist environments. Furthermore, planktonic algae and benthic algae are fishery resources, including laver, seatangle, diatoms, and so on.

5.2.2 Basic Factors and Models for Changes in Fishery Resources

5.2.2.1 Basic Factors for Changes in Fishery Resources

There are many factors for changes in fishery resources. Generally speaking, these factors can be classified as follows: the biological characteristics of fish themselves, restrictions of living environmental factors, human fishing, and so on. The fish factors include reproduction, growth, and mortality; environmental factors include water temperature, salinity, prey organisms, interspecies relationships, other hostile organisms, and so on. Many factors contribute to the quantitative change of fishery resources, and they are complex. The quantitative changes of fishery resources are often the result of a comprehensive effect of various factors combined; that is, the result of mutual restrictions between internal and external factors.

Fish reproduction is restricted by the fertility of the parent population, the fertilization rate of the eggs produced, and the survival rate of the fish eggs and larvae. Moreover, the survival rate of fish eggs and larvae is closely connected with environmental factors. First, fish growth is influenced by the population density, composition of the population at different ages, and the prey and hydrological conditions of the external environment. Second, fish mortality is classified into natural mortality and fishing, and natural mortality results from predators, disease, and rapid changes in the external environment. Additionally, population changes are also influenced by intraspecies and interspecies relationships. For example, when the prey condition is poor, some fish feed on their own eggs and larvae, and different species compete for the same prey. Thus, changes in food supply change a population.

Fishing is also one of the main factors affecting fish populations. On the one hand, with appropriate fishing, a reduced population could be compensated by the recruitment of the population; on the other hand, due to a lack of proper compensation, excessive fishing breaks the balance, leading to a significant decline in the fish population, which is so-called damage to fishery resources and unsustainable utilization, and fish stocks with slow growth rates, late sexual maturity, and long life spans are the most significant victims.

Different fishery resources may be affected by different factors. However, as for a species, it suffers from multiple and complicated factors. Generally speaking, an imbalance between population recruitment and reduction results in a change in population. Two reasons lead to an imbalance between population recruitment and reduction, and they can be summarized as follows: the natural factor and the artificial factor. There are many subfactors of the natural factor and artificial factor. Generally speaking, the factors affecting recruitment are much more complicated than the factors affecting reduction. Whereas good knowledge of factors for population change is the key to good knowledge of resources change and the situation of sustainable use, without knowledge of population change factors, we cannot have knowledge of the utilization situation of fishery resources.

5.2.2.2 Basic Models for Changes in Fishery Resources

Russell (1931) made a systematic theoretical summary of the research on the quantitative changes of fishery resources, and indicated that catches increase as fishing efforts increase within a certain limit, but after the ceiling, the more fishing efforts, the fewer catches. Based on the four factors of population growth and reduction, Russell put forward a basic model for the quantitative changes of fishery resources.

As a result of natural mortality and fishing, fish stocks are reduced, and a stock can then be compensated through the growth of juvenile fish to a fishable size, as well as the growth of existing fish. In the absence of fisheries (or in the case of undeveloped fisheries), a fish stock reduced by natural mortality could be compensated through recruitment and the growth of a juvenile fish population; that is, fish acquire resource growth through recruitment and growth, compensating for reductions caused by natural death, thus achieving a balance. With fisheries development, fishery resources suffer from a loss in quantity; meanwhile, target fish tend to have a lower age. In this case, when catches increase to the maximum allowable scale, a new balance could still be maintained, thanks to one factor or the combined factors of natural mortality, growth, and recruitment.

Four factors influence fish stocks, and they are natural mortality, growth, fishing, and recruitment. The interaction among these four factors and the changes in a fish stock are illustrated in Fig. 5.1. The basic model of fish stock changes proposed by Russell is shown in the following equation:

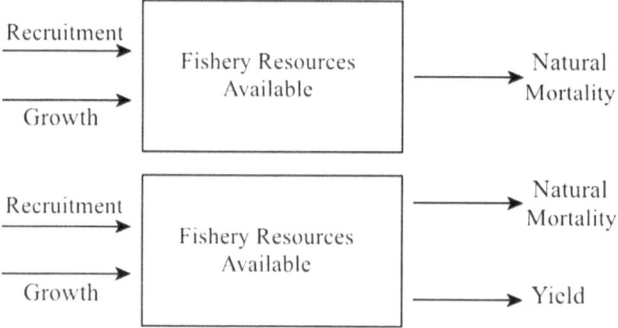

Fig. 5.1 Changes in available fish stock

$$B_2 = B_1 + R + G - M - Y \tag{5.1}$$

In this equation, B_1 and B_2 represent the resources biomass available at the beginning and end of a certain period, respectively; R, G, M, and Y represent the recruitment, growth, natural mortality, and yield (the catches), respectively.

The above equation shows that when $Y < (R + G-M)$, the fish stock increases; that is, $B_2 > B_1$; when $Y > (R + G-M)$, the fish stock decreases; that is, $B_2 < B_1$; when $Y = (R + G-M)$, then the fish stock remains balanced; that is, $B_2 = B_1$.

The upper figure shows the situation in the absence of fisheries; the lower figure shows the situation in the case of fisheries.

5.2.3 Impact of Changes in Recruitment and Surplus Stocks on the Sustainable Utilization of Fishery Resources

Fishery resources are composed of two stocks, that is, recruitment stock and surplus stock. In general, fish have inherent living habits; that is, when juvenile fish grow to a certain size, they make a feeding and overwintering migration, and enter enter the spawning ground with the original adult fish. As for the fish stock whose juveniles grow into adults within 1 year, there are no original adult fish. The fish stocks are recruited by the recruitment stock of that year, and there is no surplus stock for the fish stock.

Once juveniles grow to a certain size, they enter the fishing grounds and encounter fishing gear for the first time. From the perspective of fishery resource exploitation, a fish stock that is caught in great quantity is called surplus stock. The means by which the recruitment stock is recruited are complex, but they can be classified into three basic types: (1) one-off recruitment, (2) batch recruitment, and (3) continuous recruitment.

The reasons for changed recruitment stock are complex, while change plays a decisive role in the ups and downs of fisheries as well as the sustainable utilization of fishery resources. Although scholars hold different views for the reasons of change in recruitment stocks, these reasons could be summarized as water temperature, feed, marine currents, and so on, while the impact of these factors varies among different fish stocks. Of course, some factors are listed as the primary reasons for a changed recruitment stock, namely, the number of spawning broodstock, the fecundity and number of fertilized eggs, hydrological environment (including water temperature, current and so on), meteorological conditions, and feed foundations, and these latter three factors have important impacts on the development, growth, and survival rate of eggs and larvae. Moreover, the number of spawning broodstock is largely restricted by fishing. The number of recruitment stock is restricted by feeding availability and predation of predatory fish and other predators. Moreover, fish growth, which directly affects the average age and span of sexual maturity, has an impact on the capacity of a recruitment stock.

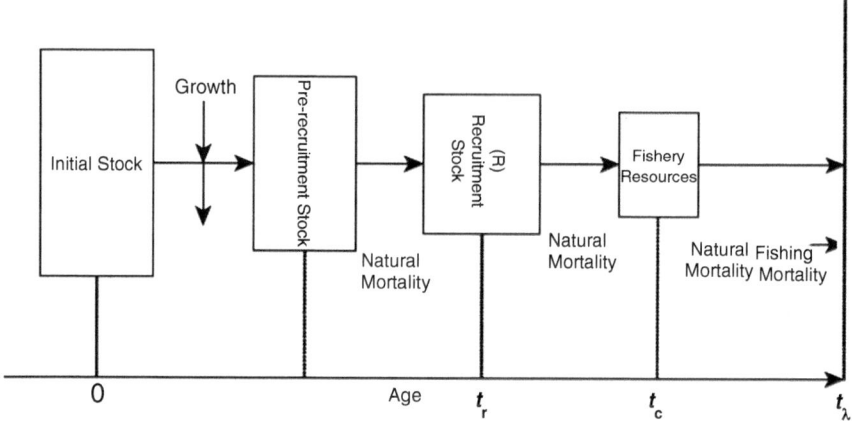

Fig. 5.2 Diagram of process of decreasing fish stock number in one generation (Chen 2014)

Fishing has the most significant impact on the amount of surplus stock, while the extent of the impact varies with the fishing intensity (including the quantity and performance of fishing gear, fishing technology, the amount of fishing operation time, etc.). In addition, the amount of recruitment stock of one generation exploited in fisheries also has an impact on the surplus stock. Generally speaking, the higher the amount of recruitment stock from one generation, the longer it is exploited in fisheries. In other words, the catch from the recruitment stock of the last generation is proportional to the amount of recruitment stock from the present generation.

Figure 5.2 is a diagram that shows the process of the decreased number of fish stock from one generation. Generally speaking, there is a great abundance of fish stock at the initial stage of a generation (Chen 2014). A fish stock reaches the age at recruitment (t_r) through natural mortality and growth, and then the fish stock is the recruitment stock (R). The recruitment stock that reaches the fishing grounds is still growing in size while decreasing in number due to natural mortality. The size of the fishing target is determined by the mesh size of a fishing net. If the mesh size is relatively large, the recruitment stock continues to decrease along with the adults and natural mortality. After reaching the age at first capture (t_c), the number of recruitment stocks is expressed with the symbol R'. Henceforth, although individuals gradually grow in size with growth, due to both natural mortality and fishing mortality, the number of individuals decreases, and individuals eventually reach the end of life (t_λ).

As analyzed above, the sustainable utilization of fishery resources is a complex system in which both human and natural environmental factors play important roles. The former is controllable and manageable, but the latter is uncontrollable and uncertain, which has made it difficult to assess the sustainable utilization of fishery resources.

5.2.4 General Process of Fisheries Resource Exploitation and Utilization and Reasons for Fisheries Resource Exhaustion

5.2.4.1 General Process of Fisheries Resource Exploitation and Utilization

Generally speaking, the process of fisheries resource exploitation and utilization has gone through four stages, namely, "underexploitation, accelerated growth in exploitation, overexploitation, and resource management" (Fig. 5.3). (1) *Underexploitation stage:* If fishery resources are not utilized or the catch is far below the potential reproduction capacity of a fish stock, the catch is at a lower level, the catch per unit effort (CPUE) is relatively high, and the marginal returns (or marginal productivity) are at a progressively increasing stage. (2) *Accelerated growth in the exploitation stage:* As fisheries become profitable, rapid development occurs, and the fishing capacity and catch rapidly increase. Furthermore, the catch per unit effort first increases and then declines. Fisheries are at a productive stage. (3) *Overexploitation stage:* If the utilization of fishery resources is continuously intensified, we enter the stage of overexploitation. Although the fishing capacity remains at a high level that exceeds the natural growth rate of fishery resources, the catch falls sharply and remains at a low level. If left unprotected, the fishery resources would collapse and be exhausted. (4) *Resource management stage:* If we realize the seriousness of the problem and strengthen the scientific management of fishery resources, the fishing

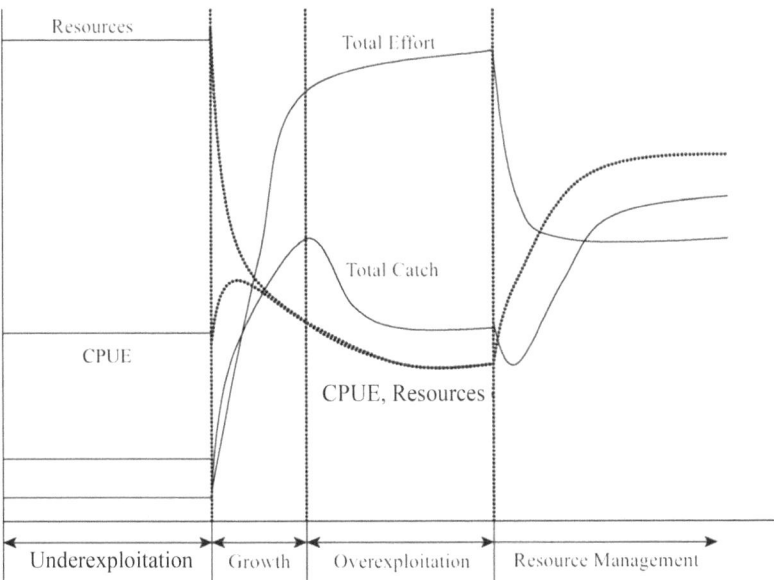

Fig. 5.3 Diagram of the different stages of typical fisheries development (Chen 2014)

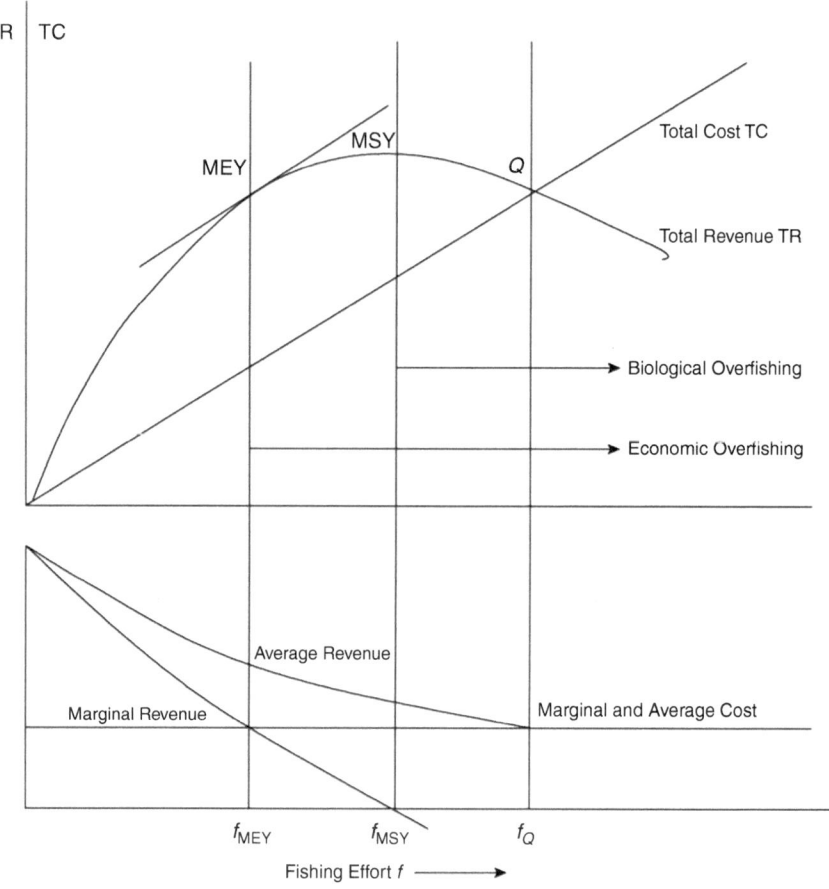

Fig. 5.4 Diagram of the economic analysis of fisheries resource exhaustion (Chen 2014)

capacity is reduced to a certain level, so that resources can be recovered and the catch increases and remains at a high level. However, in practice, only a few fishery resources can reach the fourth stage (the successful stage of management), while most traditional economic resources remain at the third stage, such as China's inshore fishery resources.

From an economic–biological perspective, since fishery resources are "shared resources," fisheries development, driven by economic benefits, is bound to pursue the breakeven point (Q point in Fig. 5.4), and the fishing effort increases to the bioeconomic equilibrium point f_Q (the fishing effort corresponding to the Q point in Fig. 5.4). Fisheries investment has inertia, so most fisheries equipment and labor forces have exceeded capacity. Meanwhile, because fisheries investment tends to be irreversible and fishing labor is segregated from other sectors, the fishing effort may exceed the level of f_Q and it is difficult for it to fall to a lower level.

As Fig. 5.4 shows, when the marginal cost of each additional unit of fishing effort equals the marginal income, the total resources rent maximizes and the catch reaches the "Maximum Economic Yield" (MEY). Whereas, from an economic perspective, there is overfishing for each additional fishing effort, that is, the marginal cost is greater than the marginal income, and the total resources rent is declining. However, the average income is still greater than the average cost, so fisheries are still profitable at this time, and the total catch may continue to increase. When the marginal income generated by catch per unit effort is zero, the total catch reaches the maximum sustainable yield (MSY). Afterward, with increased fishing effort, the total catch begins to decline, and then the marginal income becomes negative, and the total income begins to decline. From a biological perspective, this is overfishing. However, when overfishing occurs, the average income is still greater than the average cost, so fishermen can still receive a certain amount of resources rent. Therefore, the fishing effort continues to increase until the average income is equal to the average cost and there is no more breakeven point. Meanwhile, the fisheries do not produce resources rent, and the resources are severely overfished and exhausted from a biological perspective. Worse yet, some traditional fish stocks with high economic value decline while the market demand rises, which, in turn, raises fish prices, thus leading to the shift of Q to the upper right, allowing fishermen to earn high profits. A disastrous situation would further increase fishing efforts and eventually intensify the decline of fishery resources until the resources are depleted.

5.2.4.2 Reasons for Fisheries Resource Exhaustion

The sustainable utilization of fishery resources is the basis for the sustainable development of the fisheries economy. There are various reasons for fisheries resource exhaustion. Fishery resources are renewable, limited, mobile, and variable natural resources that are affected by environmental degradation, natural conditions, politics, and more. A detailed analysis of the reasons for fisheries resource exhaustion will facilitate sustainable resource utilization and a better evaluation of sustainable utilization. The main reasons for fisheries resource exhaustion are listed below (Chen 2004).

1. The contradiction between limited fishery resources and increased human demand is becoming increasingly prominent. Although fishery resources are renewable, they have a limited capability to regenerate. Before the nineteenth century, due to the relatively small population, limited technology and production capacity, people had a low demand for fishery resources, and the demand did not exceed the renewable capacity of fishery resources, so sustainable utilization of fishery resources could be achieved. Therefore, people used to believe that fishery resources were inexhaustible natural resources, and people did not change their mentality on fishery resources until the 1970s. With the development of society and an increased population, people further increased the demand for protein. Therefore, the fishing capacity greatly exceeded the renewable capacity of

resources, and the degree of intervention on natural resources was tremendously strengthened. Eventually, there was an indication that fishery resources were being exhausted. The FAO estimates that the average consumption of marine and inland aquatic products worldwide will be reduced from 10.2 kg per person in 1993 to 5.1–7.6 kg per person in 2050, which seriously threatens the life of nearly 1 billion people, most of whom are from developing countries and rely on fish as a major source of protein.

2. Economic values vary among different aquatic products, ranging from low-value species used for fish feed to high-value species consumed as luxuries. In the process of fisheries development, people have a preference for high-value species. When the demand for preferred species reaches the ceiling, the price increases sharply (many high-value species have a high price elasticity coefficient), and consumers turn to low-value species with low preference. This development model has led to the exhaustion and scarcity of many fishery resources with high economic value in the world.

3. Open and free access to fisheries accelerates fisheries resource exhaustion, which is common for many fisheries. Because marine fishery resources are mobile, it is difficult to specifically allocate exclusive utilization to specific organizations and individuals, so these resources are inevitably shared. Fishing freedom in the high seas is one of the universally accepted principles in the traditional law of the sea. Throughout the ages, the utilization of fishery resources is not entirely subject to ownership, so fishermen have complete freedom of access to fisheries and withdrawal from fisheries. In this case, fisheries are open to whoever is willing to invest in fishing gear and other equipments. As a result, fishermen are driven by short-term profits, new technologies, and the market and are invested in capturing more fish, such as through fisheries mechanization and increased fishing vessel size and horsepower. In addition, fishery resources and fishing grounds are open to the public. As a result, fishing reduces the number of fish stocks. Resource benefiters make profits from fisheries and then transfer adverse economic effects to others. This transformation is beyond the control of the market, while the benefiters do not bear the economic costs. Consequently, benefiters overexploit and overutilize resources, which has a direct negative impact on the sustainable utilization of fishery resources, because the cost borne by fishermen is not equal to the social costs. However, fishermen determine their utilization levels based on the average cost borne by them, while they do not need to take responsibility for the higher marginal social cost caused by their overexploitation and overutilization of fishery resources, leaving all users of the fishery resources to take responsibility. From an economic perspective, this is market asymmetry.

4. Uncompensated use of fishery resources is also one of the reasons for fisheries resource exhaustion. The existence of ownership should be reflected economically. If not, ownership is illusory and denies reality. The free exploitation and utilization of fishery resources leads to a disaster, which prioritizes the exploitation and utilization of fishery resources instead of conservation and management. As a result, fisheries are conducted in an extensive mode for short-term profits. With overfishing and serious waste, fishery resources are exploited in a

destructive way. According to FAO statistics, the annual abandoned bycatch accounts for approximately 32% of the world's marine fishing production (27 million tons). First, the uncompensated use of fishery resources leads to their poorly integrated utilization, poor economic benefits, and unreasonable allocation. Second, uncompensated use of fishery resources also makes it difficult to use economic measures to strengthen the management and conservation of resources. Thus, the rate of resource depletion and shortage degree cannot be accurately reflected by price signals. Meanwhile, low-value, low-age, and small-sized fish are often caught to maintain high production, while the proportion of high-value and large fish production is very low, resulting in "resources depletion" with the weakening of resource conditions.

5. Asymmetry of technological development. Asymmetry of technological progress refers to the disproportion between resource development and utilization technology and environmental protection technology. On the one hand, technological development (including the application of hybrid technology to fishing vessels, etc.) has promoted the development and utilization of fishery resources; on the other hand, technological development often neglects the protection and sustainability of environmental resources; thus, a great quantity of bycatch is caught in fishing operations. However, most of the technological advances in fisheries come from the external fisheries sector, thus making it hard to manage advances in fisheries technology.

Fisheries production has negative impacts on resource conservation. These impacts can be reduced but cannot be avoided. These impacts can be classified into different categories, and they are listed as follows: reduced resources, declined fecundity, and decreased reproductive capability; enhanced variability of the biological ecosystem; pollution of coastal and marine areas from catch processing; degradation of important habitats (by bottom trawls, large drift nets in the high seas, etc.); and environmental damage from large-scale and high-density aquaculture. Although some bycatch species have low economic value, they play an important role in the food chain and are prey for other important fish; these species also have significant impacts on the recovery and enhancement of the ecosystem and the improvement of productivity. Due to the development of other industries or agriculture, the important habitats of fishery resources in coastal waters are being destroyed, which will affect the potential fisheries resource development.

6. Surplus fisheries subsidies. As a financial and economic means, subsidies play a significant role in fisheries development and management and have a positive effect on the stability of fishing communities and the employment of fishermen. However, fisheries subsidies have played a negative role in those fisheries that are already unprofitable, because a large amount of financial subsidies motivates fishermen to keep developing and producing, giving rise to a surge in fishing capacity and a recession or even depletion of resources. Economic means, such as subsidies, are a common means for fisheries sectors to promote modern fishing techniques and high sea fisheries. Whereas publicly subsidized investment projects (such as construction and innovation of pier aquaculture facilities and

fishing vessels) usually have long life cycles and rarely have opportunities to transfer to other uses. It is difficult to adjust the fishing effort in the short term when resources decline. Moreover, the fisheries sectors have finite control over public participation in fisheries investment, and they do not directly control pier investments, financial policies, and trade or investment policies. Scholars note that fisheries subsidies are one of the main reasons for the emergence and existence of overfishing capacity as well as the decline in fishery resources.

5.2.5 Concept and Connotation of Sustainable Utilization of Fishery Resources

5.2.5.1 Connotation and Definition of Sustainability

The concept of sustainability arises from an analysis on the utilization of renewable resources, such as fishery and forestry resources. Actually, a sustainable process means that a process can be maintained endlessly for an indefinite period of time and that the quantity and quality of the resources inside or outside a system are not in decline; instead, these resources are improved. In economic terms, the exclusive use of the interest produced by the principal deposited in a bank is a sustainable process, because it keeps the amount of the principal unchanged. However, the use of interest at any higher speed destroys the principal. In terms of natural resources, the most basic and essential condition of sustainability is to keep the total stock of natural resources unchanged, or better than the current level.

The basic implication of sustainability is to meet the needs of the present generation as well as to not jeopardize future generations' capability to meet their needs, which is the sustainability of an economic–natural–social complex system concentrated in three aspects, namely, economic sustainability, ecological sustainability, and social sustainability. Economic sustainability refers to maximizing the benefits of economic development while maintaining the quality of natural resources and the services they provide; ecological sustainability refers to development that does not exceed the renewal capacity of an ecosystem, ensuring a balance between natural resources and their development and utilization; the core of social sustainability is the fair and reasonable distribution of resources among the same current generation as well as among different generations. Only with sustainable utilization can renewable fishery resources be sustainable.

5.2.5.2 Concept and Connotation of the Sustainable Utilization of Fishery Resources

Sustainable development is essentially the reasonable allocation and sustainable utilization of natural resources. In a narrow sense, the sustainable use of fishery resources means that the human fishing intensity does not exceed the endurance or self-renewal capacity of fishery resources; in a broad sense, the sustainable utiliza-tion of fishery resources means meeting the needs of the contemporary generation for aquatic products without interfering with the needs of future generations. The sustainable utilization of fishery resources should include the following aspects (Chen 2004).

1. The sustainable utilization of fishery resources must be subject to the demand of economic development for fishery resources. The ultimate goal of human pro-duction is economic development to raise the welfare level of all human beings. To some degree, economic development will inevitably be achieved at the expense of fisheries resource depletion, the pace of which is accelerated with the acceleration of economic growth. However, there is no doubt that maintaining the environmental base of fishery resources at the expense of economic develop-ment violates human desires and ethical foundations. Therefore, to ensure the demand of economic development for fishery resources, human beings can only change the way of fisheries resource utilization to achieve sustainable utilization of fishery resources to coordinate the contradiction between economic develop-ment and environmental protection of fishery resources.

2. The sustainable utilization of fishery resources refers to the whole process of development, utilization, management, and conservation of fishery resources instead of solely the use of fishery resources. Seeking and choosing the best use targets and approaches for fishery resources could give rein to the advantages and maximum structural functions of fishery resources; while "governance" is taking comprehensive measures to transform unfavorable conditions for sustainable use of fishery resources into favorable conditions, such as improving fishing ground conditions and building artificial farms, "conservation" is protecting fishery resources and the environment that used to have positive impacts on production and life. Furthermore, in regard to the utilization of fishery resources, human beings are reinvesting in the production of fishery resources rather than claiming.

3. The maintenance and improvement of the ecological quality of fishery resources is an important manifestation of the sustainable utilization of fishery resources. Fishery resource development and utilization used to, along with profits, bring about degraded ecological quality and depleted fishery resources, resulting in interference in the human demand for aquatic products, which is a lesson that is considered when a requirement is set for the maintenance and improvement of the ecological quality of fishery resources. The sustainable utilization of fishery resources means maintaining and rationally improving the fisheries resource base, as well as giving attention and consideration to ecological and environmen-tal quality when the plans or policies for fisheries resource development and utilization are made.

4. The sustainable utilization of fishery resources means the demand for a certain number of fishery resources under certain socioeconomic and technological conditions. Within the predicable prospects visible to human beings, the sustain-able utilization of fishery resources concerns fairness. To date, the utilization of fishery resources has led to a reduced amount of fishery resources, having

negative impacts on the needs of future generations; hence, this approach is not sustainable. The sustainable utilization of fishery resources must consider the needs of future generations for production and life within predictable economic, social, and technological levels.

5. The sustainable utilization of fishery resources is a comprehensive concept with consideration of society, culture, and technology rather than a simple issue relevant to the economy. With the combined effects of the above factors, human beings form a utilization mode for fishery resources under specific historical conditions. To realize the sustainable utilization of fishery resources, it is necessary to comprehensively analyze and evaluate various factors, such as the economy, society, culture, and technology, and maintain whatever is beneficial to the sustainable utilization of fishery resources, while transforming whatever is negative.

5.2.6 Factors Influencing the Sustainable Utilization of Fishery Resources

5.2.6.1 Resource Abundance and Environmental Capacity

The primary factors affecting the utilization of fishery resources in a certain region are resource abundance and environmental capacity. In economic analysis, fishery resources and the environment can be viewed as the capital necessary for production activities, and they are also the products and services that nature provides to human beings. In some sense, fishery resources and the environment are actually the ecological capital necessary for production and human life. Nonsustainable utilization of ecological capital can seriously damage fishery resources and the environment. However, the minimum safety standard establishes a boundary, which is determined by fishery resources and environmental conditions and is used to indicate the allowable level that fishery resources can be exploited. Therefore, in a specific region, the abundance of fishery resources and the size of environmental capacity directly affect the establishment of regional, minimum safety standards in the development and utilization of fishery resources, and these factors further determine the degree of achieving the sustainable utilization of fishery resources. Generally speaking, it is easier to achieve the sustainable utilization of fishery resources in areas with good fishery resources and environmental conditions than in areas with poor conditions.

5.2.6.2 Population and Economy

The influence of the population and economy on the sustainable utilization of fishery resources is mainly determined by the pressure from the population and economic development of fishery resources and the environment. On the one hand, the more fishermen, the greater the demand imposed on fishery resources and the environment, which contributes to an external environment unfavorable to the sustainable utilization of fishery resources. In addition, it is easier to break the minimum safety standard, resulting in the predatory utilization of fishery resources. On the other hand, human quality is closely related to the utilization mode of fishery resources. That is to say, the higher the human quality, the easier it is to accommodate and implement the sustainable utilization of fishery resources in consciousness and in action. There is also a strong correlation between economic development and the utilization of fishery resources. Generally speaking, the better the economic development and the greater the demand for fishery resources and the environment, the greater the potential losses to fishery resources and the environment. However, economic development provides advanced technological means for the sustainable utilization of fishery resources. Furthermore, financial support is beneficial to the sustainable utilization of fishery resources.

5.2.6.3 Technological Progress and Structural Change

Science and technology have great and unbelievable power in changing human destiny. For example, as for the marine fishing industry, the application of powered fishing vessels, new fishing materials, and navigation instruments has greatly improved the fishing capacity and intensity. Currently, when facing the problems of fisheries resource decline and environmental degradation, human beings must ensure the sustainable development of fisheries by relying on science and technology advancement. As for the development and utilization of fishery resources, we should replace technologies that have potential and realistic harm to the fisheries resource environment with technologies that are harmless or even beneficial to the fisheries resource environment. Environmentally friendly fishing gear and methods greatly reduce the environmental and ecological risks in the utilization of fishery resources. In fact, production practices can benefit from science and technology. The development and application of science and technology not only promote economic development but also play a role in reducing pollution and improving environmental quality. For example, research on fishing gear and marine farming has been applied to fisheries and promotes sustainable fisheries. Industrial, agricultural, and service industries and their internal industries have different levels of dependence on fishery resources in the economic structure of a country or region. Meanwhile, the utilization of fishery resources by various industrial sectors has different impacts on fishery resources and the environment. Therefore, effective and rational changes in the economic structure can fundamentally promote the function of the economic structure in fostering the utilization of fishery resources in a safe and preventive manner. Resource-saving and environmentally friendly industries facilitate fundamental changes in the economic structure.

5.2.6.4 Culture and Institutions

Any fisheries resource development activity is conducted under certain cultural background and institutional conditions; thus, the external constraints from culture and institutions have an important influence on the development and utilization of fishery resources. As for a national traditional culture, whether the culture contains a basic and simple mentality for fishery resources and environmental protection or not has an influence on stimulating people's internal strength and then applying sustainable fisheries resource utilization. However, institutions mainly manifest as external formal constraints. In the process of fisheries resource utilization, nonsustainable utilization behaviors of human beings cannot be effectively corrected in other forms. Therefore, the sustainable utilization of fishery resources can only be guaranteed by consciously building institutions capable of facilitating the sustainable utilization in which there are policies on paid use, property rights, prices, and more.

5.2.7 Goals for the Sustainable Utilization of Fishery Resources

5.2.7.1 Unification of Economic and Ecological Benefits

The economic benefits of fisheries resource development are a certain number of social and economic achievements targeted by economic entities. Human beings engage in production activities and produce consumption goods (basic necessities for clothing, food, shelter and transportation), which are economic products produced by an economic system. The ultimate goal of the development and utilization of fishery resources is to acquire and enjoy these economic products. However, the development and utilization of fishery resources is not only an economic process but also an ecological process. Further development and utilization of fishery resources means a change in the original ecological environment. Compared with the development mode of fishery resources in the past, the sustainable utilization of fishery resources emphasizes the harmonious unification of economic and ecological benefits in the development and utilization of fishery resources, which not only produces optimal economic benefits but also maintains and enhances the ecological quality of fishery resources. The ecological benefits mainly refer to the improved quality and increased quantity of fisheries ecological achievements that meet human ecological needs in a certain ecosystem in which human beings are regarded as the subject in the development and utilization of fishery resources.

The ecological and economic benefits are a harmonious unity formed by mutual interaction and mutual restriction, and this unity can be called the ecological economic benefits. The goal of the sustainable utilization of fishery resources is the high degree of unification of economic and ecological benefits.

5.2.7.2 Unification of Short-Term and Long-Term Benefits

In the development and utilization of fishery resources, short-sighted human beings tend to pursue short-term benefits within a certain space-time range. For long-term benefits, due to a long time span, the relevant factors are more diverse. The reasons are very complicated and, on most occasions, overlooked. Humans do not realize how large and irreparable losses are until a situation develops to a very dangerous level.

The advantages of short-term benefits are obvious during the process of development and utilization of fishery resources, so in the short term, depleted fishery resources can really increase the interest of fishermen. However, the disadvantages come after the advantages. Generally speaking, this process always takes a long time, and the losses caused are often irreversible when the disadvantages arrive.

Therefore, the goal of the sustainable utilization of fishery resources is the harmonious unification of current and long-term benefits, which is a fundamental issue for intergenerational equity. In pursuit of this unification, human beings should enhance the scientific understanding of "advantages" and "disadvantages" for the current benefits and effectively improve the predictability of long-term benefits.

5.2.7.3 Unification of Local and Global Benefits

In the process of the development and utilization of fishery resources, it is necessary to facilitate the unification of local and overall benefits, which is the key to intragenerational equity. Some fisheries resource development modes may contribute to local benefits; however, this contribution may hinder the global benefits, because the external cost of the development and utilization of fishery resources are transferred to surrounding areas. Furthermore, under some specific circumstances, in pursuit of the global benefits, the local benefits may be either temporarily impeded or reduced. The goal of the sustainable utilization of fishery resources is the unification of local and global benefits in the development and utilization of fishery resources. Human beings should, with visional arrangement, give full play to the unification between the local and global benefits to dynamically integrate and coordinate the development of local and overall benefits.

5.3 International Action for Sustainable Fisheries Development—Blue Growth

5.3.1 Overview of the 2030 Agenda for Sustainable Development

During the UN Sustainable Development Summit in September 2015, leaders of the member states of the United Nations adopted the *2030 Agenda for Sustainable Development*, which includes a set of 17 "sustainable development goals." First of all, the *2030 Agenda for Sustainable Development* defines the priorities for global sustainable development and expectations in 2030, striving to mobilize global efforts to create partnerships between human beings and the planet, realizing prosperity and peace. The agenda not only includes the "sustainable development goals" but also the *Addis Ababa Action Agenda* on financing for development and the *Paris Agreement* on climate change. "Sustainable development goals" note that the following goals should be achieved by 2030: termination of poverty and hunger; further agricultural development; support for economic development and employment; restoration and sustainable management of natural resources and biodiversity; fights against inequality and injustice; and mitigation of climate change impacts. In addition, the "sustainable development goals" are transformative and closely interactive, appealing for creative combinations of policies, plans, partnerships and investments to achieve common goals (FAO 2016).

Moreover, the FAO emphasizes that the key to achieving the *2030 Agenda for Sustainable Development* is food and agriculture. In fact, the FAO mission has contributed to the "sustainable development goals". Both the FAO "sustainable development goals" and the "Strategic Framework" make efforts to solve the problems of poverty and hunger, thereby creating a more equitable society in which everyone is taken care of. More specifically, "Sustainable Development Goal 1" (eliminating poverty in all its forms everywhere) and "Sustainable Development Goal 2" (eliminating hunger, achieving food security and improved nutrition levels and promoting sustainable agriculture) reflect the FAO vision and mission. There are other "sustainable development goals," such as gender equality ("Sustainable Development Goal 5"); the availability and sustainable management of water ("Sustainable Development Goal 6"); sustainable economic growth, full employment and decent work ("Sustainable Development Goal 8"); reduced inequalities ("Sustainable Development Goal 10"); responsible consumption and production ("Sustainable Development Goal 12"); combating climate change ("Sustainable Development Goal 13"); sustainable use of oceans, seas and marine resources ("Sustainable Development Goal 14"); sustainable use of terrestrial ecosystems ("Sustainable Development Goal 15"); and peace and justice ("Sustainable Development Goal 16"). All of the above goals are closely related. The implementation means and global partnership proposed by the parties ("Sustainable Development Goal 17") provide all FAO sectors with a basis for implementing the *2030 Agenda for Sustainable Development*, and these sectors deal with issues related to fisheries, aquaculture, and postharvest aquatic products' processing (FAO 2016).

The framework, process, stakeholder engagements, and partner relationships proposed in the *2030 Agenda for Sustainable Development* have the following merits: (1) it allows contemporary and future generations to benefit from aquatic resources and (2) it helps fisheries and aquaculture provide nutritious food for growing populations and promote economic prosperity, employment opportunities, and human well-being.

5.3.2   2030 Agenda for Sustainable Development *and Sustainable Fisheries Development*

At present, the international community has generally recognized the important role in sustainable development played by oceans, coastal areas, rivers, lakes, and wetlands, including resources and ecosystems related to fisheries and aquaculture, which was recognized at the Rio Summit in 1992 and fully embodied in Chaps. 17, Chap. 14, and Chap. 18 of *Agenda 21* as well as the *Code of Conduct for Responsible Fisheries* in 1995. In addition, the outcome document of the Rio+20 Summit also advocated for sustainable development, calling for "treating sustainable development in a comprehensive and integrated way, facilitating the harmonious coexistence of human beings and nature, striving to restore the health and integrity of the earth's ecosystems."

A number of "sustainable development goals" are linked to the sectors of capture and aquaculture, as well as to sustainable development of the sectors of capture and aquaculture. One of the goals ("Blue Target") focuses directly on the oceans ("Sustainable Development Goal 14") by highlighting the goal to "conserve and sustainably use the oceans, seas and marine resources for sustainable development," emphasizing the importance of conservation and sustainable utilization of the oceans and associated resources for sustainable development by poverty reduction, sustainable economic growth, food security, and sustainable livelihoods and decent work.

To facilitate the continued contribution of oceans and marine resources to human well-being, "Sustainable Development Goal 14" highlights the necessity to manage and sustain ocean resources, as well as support ecosystem services, which are vital to human beings. More efficient resource utilization, changes in production and consumption modes, management of human activities and improved regulations mitigate negative environmental impacts and enable contemporary and future generations to benefit from aquatic ecosystems. Therefore, the promotion of sustainable capture and aquaculture not only facilitates the management and maintenance of resources and ecological systems but also ensures that the oceans have the potential for nutritious food supplies.

Marine and inland waters greatly contribute to global food and nutrition security, livelihoods and economic growth and also provide valuable ecosystem products and services for human beings. In the atmosphere, nearly 50% of the carbon in a natural system is circulated into oceans and inland waters, but oceans and inland waters are under various threats, such as overexploitation, pollution, biodiversity loss, the introduction of invasive species, climate change and acidification. In consequence, human activities have exerted pressure on the marine bio-support system, which makes sustainable development impossible.

Approximately 31% of the commercial marine aquatic species assessed in 2016 turned out to be overfished. Mangroves, salt flats, and seagrass beds have been destroyed at an alarming rate, thus exacerbating climate change and global warming. Water pollution and habitat degradation continue to threaten capture and aquaculture-related resources in inland and marine waters. Meanwhile, people who rely on fisheries and aquaculture for their livelihoods, food, and nutrition are also at risk. Furthermore, the important contributions by fisheries and aquaculture to the well-being and prosperity of the world are being weakened by factors such as poor governance, mismanagement, and inadequate measures, while illegal, unreported, and unregulated fishing is a serious barrier to sustainable fisheries.

Some of the concrete objectives in "Sustainable Development Goal 14" call for actions from fisheries sectors. Some actions are listed below: effective regulation of fishing activities; termination of overfishing and illegal, unreported and unregulated fishing; settlement of fisheries subsidies; providing small-scale fishers with access to resources and markets; and implementation of the provisions of the *United Nations Convention on the Law of the Sea*. The other concrete objectives in "Sustainable Development Goal 14" include marine pollution prevention and marine and coastal ecosystem management and conservation. In conclusion, "Sustainable Development Goal 14" clearly notes that it is necessary to promote cooperation and coordination among all stakeholders to achieve sustainable fisheries management and better resource conservation to provide a framework for sustainable management and protection of marine and coastal ecosystems.

Currently, the integrated approach adopted in the sustainable management and development of fisheries and aquaculture, as advocated in the FAO "Blue Growth Initiative," aims to coordinate economic growth with the promotion of livelihoods and social equity. This approach focuses on balancing the sustainable and socioeconomic management of natural aquatic resources; it also emphasizes efficient resource utilization in capture and aquaculture, ecosystem services, trade, livelihoods, and food systems.

Finally, the *2030 Agenda for Sustainable Development* emphasizes the importance of establishing partnerships and strengthening stakeholder engagement, which is regarded as the key to successfully promoting and effectively implementing activities that support the concrete objectives of the interrelated sustainable development goals. To date, international initiatives that have been undertaken by the capture and aquaculture sectors include: (1) partnership in global climate, fisheries, and aquaculture; (2) advocacy and implementation of the *Voluntary Guidelines for Securing Sustainable Small-Scale Fisheries in the Context of Food Security and Poverty Eradication* by local, national, and international civil society organizations and governments; and (3) cooperation among national institutions and among the

FAO, International Maritime Organization, and International Labor Organization in combating illegal, unreported, and unregulated fishing and other fishing-related crimes mainly through measures including supporting the designation of national and regional action plans to combat illegal, unreported, and unregulated fishing, implementing the *Voluntary Guidelines for Flag State Performance*, setting up "fishing boat global records" and so on (FAO 2016).

5.3.3 Progress on International Action for Sustainable Development (FAO 2016)

After a round of consultations driven by members of the United Nations, the adopted Sustainable Development Goals Framework now includes 169 concrete objectives and 231 indicators for measuring and monitoring progress at the global level."Sustainable Development Goal 14" includes ten concrete objectives, some of which explicitly address fisheries-related issues, and some may also have a direct impact on fisheries. All objectives are established by the Inter-Agency and Expert Group on Sustainable Development Goals and supported by the index adopted by the United Nations Statistical Commission. Three objectives are as follows:

1. By 2020, recover fish stocks to the highest sustainable yield level that is determined by biological characteristics; recover fish stocks in the shortest time by terminating overfishing, illegal, unreported, and unregulated fishing and destructive fishing activities, as well as implement scientific management plans.
2. By 2020, prohibit certain fisheries subsidies that encourage overcapacity and overfishing; eliminate subsidies that encourage illegal, unreported, and unregulated fishing; and terminate the introduction of new subsidies the same as above; recognize that providing developing and underdeveloped countries with reasonable and effective special differential treatment should be listed as an indispensable topic for the negotiations conducted within the World Trade Organization on fisheries subsidies.
3. Provide small-scale individual fishermen with access to marine resources and markets.
   Under the "Coastal Fisheries Initiative" framework supported by the FAO/GEF (Global Environment Fund), specific actions are being taken to establish and implement a fisheries performance evaluation system for the following: effective evaluation of the influence of coastal fisheries projects; monitoring changes in the environmental, social, and economic benefits of fisheries; promotion of knowledge sharing by seeking an approach to implement management strategies for sustainable fisheries development.

5.3.4   Blue Growth

5.3.4.1   Priorities of Blue Growth

The "Blue Growth Initiative" focuses on achieving sustainable fisheries, reducing the degradation of fish habitats and conserving biodiversity, which calls for data-based assessment and monitoring natural resources status (including fishery resources, aquatic ecological systems, water and land, aquatic genetic resources, etc.) and the performance and sustainability of fisheries (FAO 2016).

Assessment and Monitoring of Fish Species

The "Blue Growth Initiative" recognizes that fishery resources are of vital importance for sustainable fisheries and that fisheries resource assessments are essential for understanding the fishery resources status. Data are essential for assessing the quantity of resources, while the problem is data shortage. However, preventive management can be assisted with various estimation methods, such as expert judgment. Data coverage and quality often have an influence on the accuracy of assessment results. In addition, assessment results often precede management actions. To solve this problem, an adaptive management method is adopted with a predetermined fishing model as the reference. This method requires high-quality data on catches, fishing efforts, and other related data, and the data should be shared by all stakeholders. Some databases (e.g., FishBase2 and SealifeBase3) have provided stakeholders with comprehensive ecological and biological data. Information technology and data management will further enhance the role of data in assessing resource status.

   In addition to another important measure in achieving more efficient fisheries management is the sharing of resource assessment results. Useful information can be available through comparative analysis of a target fish species and other species whose status has been assessed as well as comparative analysis of a target species and other species in other regions. This useful information throws light on the priorities in fisheries monitoring. Actually, "fisheries and resource monitoring systems" will push this work forward through overall assessment of a known fish species to summarize the fish stock assessment results, but more assessment results are required for the system to reach a comprehensive conclusion.

Protection of Biodiversity and Habitat Restoration

The "Blue Growth Initiative" recognizes the necessity to restore degraded habitats and conserve biodiversity to increase the productivity and sustainability of fishery systems. Meanwhile, an integrated biodiversity information base is currently being developed, which includes information on the quantity and occurrence of aquatic species, to better monitor related changes and to map diversity and ecological footprints. For instance, the "Marine Organism Geographic Information Systems," built by the efforts of taxonomists and ecologists around the world, provides us with a unique global source of information about species occurrence. In addition, multiple analysis models for species distribution mapping (AquaMaps) and the distribution and evolution of biodiversity richness are being developed to further understand changes in species range and their environmental and socioeconomic influence with global climate changes.

Furthermore, to minimize the negative effects of fishing on biodiversity (such as the iconic marine mammals by caught in tuna fishing or the sponges and corals in fragile ocean ecosystems), relevant data must be available in management strategies development. These data can be collected by individual observations of bycatch species during fishing operations or "unplanned encounters" with indicative species. In fact, data collection can be conducted by scientific observers on fishing vessels or the involvement of fishermen. However, the former method is costly and biased, while the latter has problems concerning security and privacy. Despite the enormous potential of automated systems for image recognition technology, this method may currently be difficult to apply widely.

Combating Illegal, Unreported, and Unregulated Fishing

The "Blue Growth Initiative" attaches great importance to combating illegal, unreported, and unregulated fishing. In this regard, the development of information technology has completely changed the method of data collection. The technologies are mainly listed as follows: fishing vessel registration and license database sharing to facilitate the evaluation of fishing authorization; automatic identification systems and fishing vessel monitoring systems to promote monitoring of fishing vessel trajectory; electronic logs to report catches in real time; on-board video surveillance for comprehensive observation of fishing activities; port use communication for law enforcement; electronic transmission of market information to improve traceability; and catch records. These technologies ensure rigorous and efficient monitoring, surveillance and control, tracking fish products through trade certification in the distribution chain, and obtaining overall statistics based on data from operators. In conclusion, it is of vital importance to promote information sharing among responsible users through globally standardized electronic monitoring, surveillance, and control, which helps eliminate coverage blind spots and prevent illegal, unreported, and unregulated fishing.

Performance Monitoring and Sustainability Enhancement

Fisheries performance embodies social economics, the environment, and management. The investigation of fish stocks is a starting point for understanding and propagandizing the socioeconomic importance of fisheries, which is concretely illustrated by people's participation, economic investment (the size and quantity of fishing vessels), and returns (catches and monetary value). To influence policy and management decision making, the FAO recommends regarding the investigation of fish stocks as a method of raising awareness on small-scale fisheries and related livelihoods. Moreover, the investigation of stocks can also be an approach to understanding the potential impacts of fisheries on biodiversity, including listing bycatch species. The investigation of aquaculture farming provides decision makers with knowledge to help their efficient planning and management. In addition, this investigation helps check the effectiveness of fisheries management in achieving sustainability. In return, this investigation affects consumer purchase behavior, which creates an incentive for management improvement, and the increasingly widespread use of fish ecology labels is a case in point.

5.3.4.2 Focus of the "Blue Growth Initiative": Maximizing Social and Economic Benefits

To achieve the maximization of social and economic benefits, it is necessary to monitor the performance and sustainability of activities related to the utilization of aquatic resources in a value chain, while this monitoring should be conducted separately from other agricultural and commercial activities. Nevertheless, the social and economic contribution information from the fisheries industry is fragmented and often merged with information from other industries, which focuses on commercial activities in the primary production sector (excluding artisanal and subsistence fisheries) and fails to cover the entire value chain or related activities. In fact, data deficiency leads to mistakes in policy making.

The "Blue Growth Initiative" focuses on assessing ecosystem services provided by ecosystems to aquatic resources. These services are provided to a wide range of fisheries sectors including recreational fisheries and fisheries-related tourism projects. These services could contribute to biodiversity and habitat and ecosystem resilience (e.g., mangroves protecting coastal communities). Additionally, these services also include climate change mitigation, such as the carbon cycle enhanced by algae and carbon sinks facilitated by mangroves or coral reefs.

5.3.4.3 International Action for Blue Growth

We have come to realize the importance of data to blue growth. For example, the European Ocean Commission has urged that funding for European public research should go to basic research on little-known, deep-sea systems and the establishment of environmental benchmarks. Another example is that the Ocean Ecosystem Action Plan for the continental shelf in the Caribbean and northern Brazil aims to eliminate the destructive threats blocking the blue growth process in this region. One of the assistance plans emphasizes arrangements related to governance and cooperation, as well as promotes collaboration among multiple independent initiatives on prevention of habitat degradation, unsustainable fisheries, and pollution. Meanwhile, this program will summarize the information on marine ecosystems and marine living resources and make a comprehensive data table, which will be available online.

It is necessary for the "Blue Growth Initiative" to facilitate interdepartmental and value chain integration for data collection, especially for data collection on the assessment of sustainable socioeconomic benefits. First, enhanced information standards and improved coordination will facilitate information exchange, as it will drive people to use uniformed classification standards, concepts, and data structures; second, global, regional, and national data- and information-sharing platforms must be available; finally, it is important to strengthen partnerships and other cooperation arrangements, because organizations cannot meet all related requirements of the "Blue Growth Initiative" by working independently. Although the current FAO strategies remain valid and serve as a guideline, the limitations mentioned above still indicate that some areas must be listed as a priority. Therefore, the FAO calls for a global partner relationship/alliance to create a global data framework for blue growth. The FAO will coordinate among different partners and provide the necessary back-up for the collection and comprehensive utilization of data from different initiatives and disciplines (e.g., databases, information standards, methods, tools, professional capability, and collaborative data facilities).

5.4 Carbon Sink Fisheries

5.4.1 Concept and Role of Carbon Sink Fisheries

By defining a carbon sink and carbon source as well as the characteristics of marine biological carbon sequestration, carbon sink fisheries refer to the process and mechanism of promoting the absorption of CO_2 by aquatic organisms through fisheries production activities, as well as removing this carbon from waters by fishing, which is also known as a "movable carbon sink" (Tang and Liu 2016). Carbon sink fisheries are a general term for fisheries production activities that can achieve the full potential of the carbon sink function, including directly and indirectly absorbing and storing CO_2 from water and reducing the atmospheric CO_2 concentration, thus slowing down ocean acidification and climate warming (Tang and Liu 2016). Therefore, biological carbon sinks are facilitated by fisheries activities that do not involve bait casting. Thus, carbon sink fisheries have many forms, including aquaculture of algae, shellfish and filter-feeding fish, artificial reefs, enhancement and artificial release, capture fisheries, and more (Tang and Liu 2016).

Carbon sink fisheries have the following roles: improving the water's capability to absorb atmospheric CO_2; absorbing CO_2 from water by aquatic organisms (e.g., plankton and algae); and removing carbon from water by capturing aquatic products. In conclusion, the process and mechanism of carbon sinks can improve the water's capability to absorb atmospheric CO_2, thereby contributing to CO_2 emission reduction.

1. Marine aquaculture and carbon sink fisheries. The concrete forms of carbon sink fisheries are as follows: processing and production activities of aquaculture organisms (seaweed, shellfish, etc.) that absorb carbon from water through photosynthesis and consume great amounts of filter phytoplankton; consumption of carbon by biological species (e.g., fish, cephalopods, crustaceans, and echinoderms that feed on plankton, shellfish, and algae) through food web mechanisms and growth activities. Therefore, in the era of a low-carbon economy, China, as a fisheries superpower, should actively develop carbon sink fisheries, prioritizing marine aquaculture, and seize the technological heights of a blue low-carbon economy.

2. Ecological fisheries in reservoirs. Ecological fisheries in reservoirs also fall under the category of carbon sink fisheries. The carbon sink process of ecological fisheries in reservoirs is as follows: a large number of organics brought into the water through surface runoff—organics decomposed by microorganisms into inorganics, such as nitrogen and phosphorus—inorganics consumed by algae and aquatic plants accumulated as nitrogen; phosphorus, and other nutrient salts in a fish body from consuming various algae, aquatic plants, and animals; and carbon eliminated from water through fishing. Therefore, fishing in a reservoir is the most effective way to remove nitrogen and phosphorus from the water. Fisheries significantly improve the carbon sinks and water quality of reservoirs where fisheries are conducted. Within certain limits, the higher the fishing production, the more significant the effect of carbon sinks on water purification. Thus, the development of carbon sink fisheries serves multiple purposes by providing more high-quality protein for people, reducing CO_2 emissions, and relieving water eutrophication. The case in point is ecological fisheries such as shellfish and algae cultures in shallow water and ecological fisheries by releasing silver and bighead carp in reservoirs.

5.4.2 Marine Carbon Sink Fisheries and Expansion Strategy

Biological carbon sequestration is safe, efficient, economical, and feasible. In addition to carbon sequestration in terrestrial ecosystems such as forests, grasslands, and swamps, carbon sequestration in marine ecosystems has also attracted worldwide attention. Marine carbon not only directly affects the global carbon cycle through regulation and absorption but also effectively delays the impact of greenhouse gas emissions on the global climate with its enormous carbon sink function by absorbing 20%–35% of the total amount of CO_2 emitted by human beings (approximately 2×10^9 t). Overall, the ocean is the largest long-term carbon sink. According to the Blue Carbon Report released by the United Nations Environmental Programme, marine organics (including plankton, bacteria, algae, salt marshes, and mangroves) have sequestered 55% of the world's carbon. Marine plants (e.g., seagrasses, algae, mangroves, etc.) have an extremely high capacity and high efficiency for carbon sequestration. Although the biomass of marine plants is only 0.05% of terrestrial plants, their carbon reserves are almost the same. Marine biological carbon sequestration constitutes a carbon capture and removal channel, allowing biological carbon to be stored for as long a period as one thousand years, so marine biological carbon is also known as blue carbon or blue carbon sinks. Productivity for coastal areas is oriented by blue carbon sinks, providing human beings with enormous benefits (e.g., sinks are regarded as buffers for pollution and extreme weather events as well as sources of food, livelihood security, and social well-being) and services, estimated to exceed 25 trillion dollars annually. Approximately 50% of the world's fisheries are conducted in these coastal waters.

Marine carbon sink fisheries are an important component of marine blue carbon sinks. Marine carbon sink fisheries are regarded as the carbon sink activity with the most potential for expansion. Healthy, ecological, and sustainable carbon sink fisheries can be facilitated by implementing management measures such as conservation and expansion combined with conservation, restoration, and improvement of the blue carbon sequestration capacity in natural waters. Accordingly, China's marine capture and aquaculture industry is expected to achieve a blue carbon sequestration of 4.6×10^8 t/a, equivalent to approximately 10% of annual carbon emissions. Meanwhile, carbon sink fisheries also concretely manifest the new concept of green and low-carbon development in fisheries. Carbon sink fisheries demonstrate the service functions of climate regulation, water purification, and food supply. At the same time, vigorous development of carbon sink fisheries not only positively contributes to mitigating global climate change but also has vital practical significance for food security, water resource and biodiversity conservation, increased employment, and enhanced fishermen's income.

5.4.2.1 Research on Marine Carbon Sink Fisheries

Marine carbon sink fisheries include the "carbon sequestration" process of carbon absorption from seawater by cultured shellfish through filter feeding and by algae through photosynthesis; this process also includes the carbon consumed by fishing target species that feed on plankton, algae, and shellfish (e.g., fish, cephalopods, crustaceans, and echinoderms) by ingestion and growth. All the fisheries production activities that do not need bait casting function as carbon sinks; thus, they are called carbon sink fisheries. Up to now, marine fisheries have hardly attracted attention as a carbon sink industry.

Efficient Carbon Sequestration from Marine Shellfish and Algae Aquaculture

Marine plants (e.g., algae and so on) are recognized organics capable of facilitating efficient carbon sequestration. The process of how marine plants involve carbon sequestration is as follows: marine plants directly absorb CO_2 from water through photosynthesis, increasing ocean carbon sinks and promoting and accelerating the transfer of CO_2 from the atmosphere to seawater, which is conducive to reducing atmospheric CO_2. Take the role of shellfish in carbon sequestration as an example. Shellfish play a role in reducing carbon emissions, because they feed on a large amount of phytoplankton in the water during their cultivation and growth. At the same time, shellfish directly absorb the bicarbonate (HCO^-) in seawater to form calcium carbonate ($CaCO_3$) during shell formation. One mol of carbon can be absorbed in the formation of 1 mol of calcium carbonate. During the growth cycle, 30% of oceanic carbon absorbed by a scallop is removed from the waters by harvesting and 40% sinks to the seabed (most is sequestrated in the seabed). Furthermore, the carbon sequestration rate of scallops cultured at Sanggou Bay of Shandong Province is calculated at 3.36 tC/(hm^2 a), which is not only significantly higher than that of blue carbon organics in natural waters but is also higher than plantations in China over the past 50 years (1.9 tC/(hm^2 a)), reaching or slightly exceeding the annual variation limit of carbon reserves of forest biomass per unit area in developed countries or regions, such as the European Union, the USA, Japan, and New Zealand (0.25–2.60 tC/(hm^2 a)). This finding means that the "carbon sequestration" effect of seawater shellfish and algae culture is efficient and has a significant carbon sink function. The total yield of cultured seawater shellfish and algae in China from 1999 to 2008 is calculated at 8.96×10^6–13.51×10^6 t, and the average annual carbon sequestration is 3.79×10^6 t, of which 1.2×10^6 tC is removed from the seawater (excluding the sequestrated storage in the seabed). The contribution of Chinese seawater shellfish and algae cultivation to reducing atmospheric CO_2 is estimated to be equivalent to the contribution of plantations with 5×10^5 hm^2 constructed annually and continuously for 10 years (5×10^6 hm^2 by the end of the tenth year). In 2014, the production of shellfish and algae aquaculture in China was 13.17×10^6 t and 2×10^6 t, respectively. The carbon sequestration from algae cultivation was nearly 5.31×10^6 t, and the removed carbon was 1.68×10^6 t (1.17×10^6 t from shellfish aquaculture and 5.1×10^5 t from algae aquaculture). Different aquaculture modes vary in the ecosystem service value that they can offer; that is, their carbon sink efficiency also varies. The potential for carbon sinks from shellfish and algae aquaculture, whether carbon sinks as a whole or in a unit area, is likely to expand.

Other Fisheries Capable of Carbon Sinks

As mentioned above, the carbon processed in carbon sink fisheries includes not only that consumed by cultured shellfish and algae, which are at a lower tropic level, but also that consumed by certain species of living resources through feeding and growth activities. These living resources, which are at a higher tropic level, feed on natural prey such as phytoplankton, shellfish, and algae, which are at a lower trophic level. A considerable amount of carbon is removed from the water when the living resources are removed from the water by fishing and harvesting. In the view of foreign scholars, restocking whales and large-sized fish should be an effective way to improve the function of marine carbon sinks, which can even be equivalent to the construction of plantations, which is believed to mitigate climate warming by increasing primary productivity. The measurement of forest carbon sinks can serve as a standard measurement for carbon sink fisheries, which is believed to serve as a reference when a fishing quota is sold in light of carbon credits. In conclusion, attention should be paid to carbon sink fisheries, which may well be an efficient way of carbon sequestration.

5.4.2.2   Main Problems with the Expansion of Carbon Sink Fisheries

Establishing a Measurement Method for Carbon Sink Fisheries

The marine carbon cycle is the core of global carbon fluxes, and the core of studying the ocean carbon cycle is the accurate determination of various parameters. According to the Marine Carbon Advisory Group of the International Commission for Marine Research and the Intergovernmental Oceanographic Commission of the United Nations Educational, Scientific and Cultural Organization, accurate detection of 4 parameters (pH, alkalinity, dissolved organic carbon, and CO_2 partial pressure) is the key to measuring marine carbon sinks. The popular physical and biogeochemical methods for measuring marine carbon sinks include the box model, general circulation model (GCMS), on-site dissolution of organic carbon and measurement of its ^{13}C, atmospheric O_2/N_2 time series and measurement of ^{13}C, global air–sea interface carbon flux integration, etc. With the carbon flux method, some scholars have found that a sea shelf is a large container for carbon sinks and that plant carbon sequestration is very important. At present, the measurement and monitoring of carbon sink fisheries is still in the preliminary stage of experimentation, with energy ecology and box ecological models as the important measurements. However, there is still a lack of monitoring technology for accurate carbon sink fisheries measurement.

Contradiction Between Overfishing and Development of Carbon Sink Fisheries

From 1980 to 2000, the annual carbon sequestration of the Bohai Sea fishing industry was an estimated 2.83×10^6–1.008×10^7 t, and the annual carbon sequestration of the Yellow Sea fishing industry was approximately 3.61×10^6–26.13×10^6 t (Zhang et al. 2013). Generally speaking, carbon is primarily sequestrated by phytoplankton and converted into the biomass of harvested species. Therefore, increased fishing production means an increase in the amount of carbon removed from marine ecosystems. However, overfishing has weakened the function of carbon sink fisheries. The annual carbon sequestration of the Yellow Sea and Bohai Sea fishing industries has decreased by 23% and 27%, respectively. Meanwhile, the decline in resources has led to reduced carbon sequestration in waters and seabeds, which is also detrimental to the sustainable carbon sink function of the fishing industry. In addition, overfishing reduces the trophic levels of marine ecosystems, shortens the food chain, simplifies the food web structure, and miniaturizes individual species of capture fisheries, thereby reducing the contribution of capture to marine carbon sinks. Therefore, to increase the carbon sinks of marine organics, especially those related to capture fisheries, it is necessary to strictly control overfishing.

5.4.2.3   Key Technical Requirements for Expanding Carbon Sink Fisheries

Technology for Integrated Aquaculture with Multiple Trophic Levels

Shellfish and algae aquaculture and integrated aquaculture with multiple trophic levels are effective ways to cope with significant changes in inshore ecosystems under multiple stresses and to maintain carbon sinks in inshore fisheries. These ecosystem-friendly aquaculture modes not only promote efficient ecosystem production but also maximize climate regulation ecosystem services. Therefore, it is important to strive for the development of carbon sink fisheries technology for healthy, ecological, and integrated aquaculture with multiple trophic levels; to continuously optimize the modes; and to conduct systematic and comprehensive research on the carbon sink function and mechanism of these aquaculture modes (Tang and Liu 2016).

Technology for the Cultivation and Conservation of Seaweed Beds

In the global marine ecosystem, seaweed, whose distribution area accounts for less than 0.2% of the ocean, contributes 10%–18% of the total annual carbon sequestration of the oceans all over the world. Seaweed beds are the key ecological environment for fisheries organics, taking up multiple ecological roles as spawning grounds, nursing grounds, breeding grounds, etc. Therefore, seaweed beds play a very important role in marine carbon sequestration. Considering the rapid disappearance of seaweed beds worldwide, research is needed on the role of seaweed bed protection, transplantation, and cultivation in the expansion of carbon sink fisheries (Tang and Liu 2016).

Technology for Intensive Land- and Shallow Water-based Aquaculture

The development of land-based, industrialized aquaculture with a circulating water supply and pond aquaculture with a circulating water supply is an important development trend for upgrading and renovating the aquaculture industry in China. With circulating water, the enhanced efficiency by intensification and centralized collection and treatment of aquaculture waste, energy conservation, emission reductions, and ecological efficiency can be enhanced in the aquaculture industry, and the expansion of carbon sink fisheries can also be enhanced. In 2014, the total marine aquaculture farming in China occupied 2.31×10^6 hm^2, of which 0.13% is industrialized aquaculture. Production from industrialized aquaculture accounts for only 0.94% of the total amount of marine aquaculture, and industrialized aquaculture with circulating water supply only accounts for less than 50%, indicating that this specific aquaculture mode has a great potential for development (Tang and Liu 2016).

Technologies and Facilities for Deep-Sea Aquaculture

For the expansion of carbon sink fisheries, it is of great significance to expand the farms for deep-sea aquaculture and for aquaculture of species such as shellfish and algae, which are nonbait casting species, but the key to expansion is a breakthrough in the technologies and facilities. Take the case of China's deep-sea cage aquaculture as an example. With nearly a decade of development, the production of deep-sea cage aquaculture in 2014 accounted for only 0.62% of total marine aquaculture production and 17% of total cage aquaculture. In fact, the key restriction in the development of deep-sea aquaculture is the engineering facilities, which are not sufficiently adequate for long-term maintenance in the high seas. In addition, due to a lack of efficient and durable deep-water aquaculture facilities (including hoisting machines, washing machines, harvesting machinery, etc.), the problems of high risk and high-labor intensity in deep-sea aquaculture have not been fundamentally solved. Therefore, it is urgent to develop technology for deep-sea facilities and techniques for new production processes (Tang and Liu 2016).

5.4.2.4 Countermeasures and Suggestions for the Development of Carbon Sink Fisheries

Identification of the Potential and Dynamic Mechanism of Marine Carbon Sink Fisheries

To fully understand the carbon sink potential of global marine fisheries, it is necessary to establish measurement and assessment technologies for marine biological carbon sinks and carbon sink fisheries, and it is equally necessary to establish systematic, inshore ecosystem carbon fluxes and carbon sink–monitoring and observation stations. Meanwhile, it is also necessary to strengthen basic scientific research, to integrate ecological and biogeochemical research methods, to improve the existing oceanic carbon flux models, to study carbon fluxes and carbon sequestration mechanisms for major marine organics, and to assess the characteristics and dynamics of global marine fisheries, biological carbon sources, and their carbon sinks. Then, a comparative study could be conducted on the carbon sinks of different fisheries, a fisheries carbon source convergence model could be established, and the uncertainty of carbon sink estimates could be reduced.

Exploration of New Measures to Expand Carbon Sink Fisheries

- *Vigorous development of carbon sink fisheries with marine aquaculture as the priority.*

The global marine aquaculture industry is a carbon sink fishery with shellfish and algae aquaculture as the main component. As a nonbait casting fishery with a low trophic level, shellfish and algae aquaculture plays an important role not only in the supply of aquatic products, food security, and so on but also in the improvement of the ecological aquatic environment and alleviation of global warming. In short, shellfish and algae aquaculture can bring about very significant ecological, social, and economic benefits. Therefore, countries need to make strategic plans for the development of marine aquaculture to expand carbon sink fisheries by encouraging healthy, ecological, and environmentally friendly aquaculture; promoting the construction of marine ecological ranches; reducing fishing intensity; and expanding the scale of fisheries proliferation. In doing so, the carbon storage of marine carbon sink fisheries can be increased (Tang and Liu 2016).

- *Enhanced conservation and management of inshore natural carbon sinks and protection of the ecological environment.*

Mangroves, coral reefs, salt marshes, and natural seaweed beds play important roles in marine carbon sinks. Therefore, effective measures should be taken to conserve the existing marine flora. Transplantation and planting of seaweed, seagrass, and coral are still important methods to restore and expand marine blue carbon sinks. However, at present, there are still many technical problems in seaweed bed transplantation and reconstruction around the world. As a result, it is necessary to build artificial seaweed beds, strengthen conservation and management, restore the service function of marine ecosystems, and expend blue carbon sinks (Tang and Liu 2016).

- *Implementation of carbon sink fisheries expansion projects.*

First, technological and industrial demonstration projects are key in carbon sink fisheries. It is important to enhance the understanding and conduct of carbon sink fisheries with marine aquaculture as the priority, thus giving priority to the carbon sink function of fisheries organics and providing a demonstration of the development of green and low-carbon industry. Five projects are recommended for the development of carbon sink fisheries: seed enhancement for marine aquaculture proliferation, ecologically healthy aquaculture proliferation, safe and green feed supply, enhanced aquaculture facilities and equipment, enhanced technology, and equipment for deep and intensive processing of aquatic products: Of these, the main projects focus on development of integrated aquaculture with multiple trophic levels as well as technology for deep-sea aquaculture proliferation (Tang and Liu 2016).

Second is the construction and management of large-scale marine "forest grass-land" projects. It is necessary to vigorously carry out public welfare projects that aim to improve natural inshore carbon sink functions all over the world. Various measures should be taken to strengthen the conservation and management of marine organics capable of natural carbon sinks, such as the construction of seaweed beds in shallow water and large algae aquaculture in deep water and the development and utilization of new materials for biomass energy (Tang and Liu 2016).

References

Chen XJ (2004) Evaluation theory and method of sustainable utilization of fishery resources. China Agriculture Press, Beijing. (in Chinese)

Chen XJ (2014) Fisheries resources economics. China Agriculture Press, Beijing. (in Chinese)

Chen XJ, Zhou YQ (2018) Introduction to fishery. China Science Press, Beijing. (in Chinese)

FAO (2016). The State of World Fisheries and Aquaculture 2016. Rome

Russell ES (1931) Some theoretical considerations on the over fishing problem. J Cons Per Int Exl Mer 6:3–27

Tang QS, Liu H (2016) Carbon sink for marine fisheries and its expansion strategy. China Eng Sci 03:68–73. (in Chinese)

Zhang B, Sun S, Tang QS (2013) Carbon sink function of marine capture. Prog Fish Sci 01:70–74. (in Chinese)

Chapter 6
Global Environmental Change and Fisheries

Xinjun Chen

Abbreviations

DIVERSITAS	International Program of Biodiversity Science
ENSO El	Niño-Southern Oscillation
ESSP	Earth System Science Partnership
GEC	Global environmental change
GHGs	Greenhouse gases
IGBP	International Geosphere-Biosphere Program
IHDP	International Human Dimension Program on Global Environmental Change
IPCC	Intergovernmental Panel on Climate Change
SES	Social-ecological system
UNEP	United Nations Environment Programme
UV-B	Ultraviolet-B
WCRP	World Climate Research Program

Global environmental change (GEC) means a series of biological physical changes in the land, ocean, and atmosphere driven by an interwoven system of human activities and natural processes that have has tremendously threatened human beings. Because of the increasingly serious problem of global environmental change, the Earth System Science Partnership (ESSP) carries out research on the integrated study of earth systems to promote research on earth system integration and changes as well as to promote research on the capabilities of global sustainable development in light of these changes. Research on change is mainly composed of the following

X. Chen (✉)
College of Marine Sciences, Shanghai Ocean University, Lingang New City, Shanghai, China
e-mail: xjchen@shou.edu.cn

© Science Press & Springer Nature Singapore Pte Ltd. 2020
X. Chen, Y. Zhou (eds.), *Brief Introduction to Fisheries*,
https://doi.org/10.1007/978-981-15-3336-5_6

four great global environmental change programs: (1) World Climate Research Program (WCRP); (2) International Geosphere-Biosphere Program (IGBP); (3) International Human Dimension Program on Global Environmental Change (IHDP); and (4) International Program of Biodiversity Science (DIVERSITAS). Global environmental problems mainly involve global warming, ozone depletion, acid rain, eutrophication, forest deterioration and decreased biodiversity, desertification and water shortage, marine pollution, etc.

6.1 Overview of Global Environmental Change

6.1.1 Eutrophication

Pollution brought about by human activities has spread across the globe and even to remote Antarctica. According to current statistics, human beings emit nearly 10 billion tons of various waste gases into the atmosphere every year, and the total amount of industrial wastewater and domestic sewage is more than 200 million tons. As for the wastewater and sewage, except for a small portion left in rivers and lakes, the rest eventually enters the ocean. Meanwhile, most waste discharged into the atmosphere (including greenhouse gases) flows into the ocean through various channels and forms including rain, snow, air convection, and so on. Human activities cause a variety of pollution. Among these pollution types, water eutrophication is one of the important reasons for quantitative and qualitative changes in the structure of aquatic ecosystems and ultimately leads to the degradation of fisheries, among which the production of fish species with important economic value declines sharply.

Eutrophication is water pollution caused by a high content of nutrients such as nitrogen and phosphorus. Under natural conditions, nutrients are continuously deposited and sedimented at the lake bottom. With alluvial deposits and aquatic debris accumulated in a lake, the lake is transformed from an oligotrophic lake into a eutrophic lake and then evolves into a swamp and land, although this process takes a very long time. However, human beings discharge a large amount of industrial wastewater and domestic sewage. Meanwhile, plant nutrients in farmland also run off into slow-flowing waters including lakes, reservoirs, estuaries, bays, etc. Afterward, aquatic organics, especially algae, boom and bring about changes to the biological species and quantity in the water, thus destroying the aquatic ecological balance. Then, a large number of dead aquatic organics are deposited at the bottom of the lake and then decomposed by microorganisms, which consume a large amount of dissolved oxygen. As a result, the dissolved oxygen content in the water decreases drastically and the water quality deteriorates, which has a negative impact on fishes' survival and greatly accelerates the water eutrophication process.

When water eutrophication occurs, a plankton bloom develops as well, so that the water becomes blue, red, brown, milky white, and so on, which is called a phytoplankton bloom in rivers and lakes, or a red tide in the sea. In a sea where a red tide occurs, some plankton explosively blooms, and the water turns red, the reason why it is called a "red tide." Algae in a red tide are malodorous and poisonous, not edible for fish.

In addition, phytoplankton blooms also cause water anoxia, which kills aquatic animals, especially cage-cultured species or keeps fish away from the water. For instance, one of the large aquaculture farms in Japan is located in the waters around Uwajima Bay and on the eastern side of Hoketsu Bay in Japan. In the farm waters, a toxic red tide called *Gonyaulax polygramma* occurred in 1994, which dealt a heavy blow to the marine aquaculture industry there. A survey found that the reasons for the mortality of cultured species from this red tide were hypoxia and the formation of large-scale anoxic waters accompanied by high concentrations of sulfides and nitrogen.

6.1.2 Global Warming

Global warming refers to the phenomenon of climate change caused by the greenhouse effect in the earth's atmosphere and oceans over a period of time, and it is one of the most common tragedies. In simple words, the phenomenon of the earth warming is known as global warming. Over the last century, the global average temperature experienced four fluctuations (cold $\rightarrow$ warm $\rightarrow$ cold $\rightarrow$ warm), and the temperature maintained an upward trend in general. Since the 1980s, the global temperature has significantly increased.

Global warming has always been a hot topic for scientists. Many scientists believe that the primary reason for global warming is that atmospheric CO_2 emissions have increased. According to a survey by the International Energy Agency, the USA, China, Russia, and Japan account for almost half of the total global CO_2 emissions. In addition, this survey indicates that US CO_2 emissions rank the first in the world, with annual per capita CO_2 emissions of approximately 20 tons and 23.7% of the global emissions. Whereas, the annual per capita CO_2 emissions of China are approximately 2.51 tons, and the overall CO_2 emissions account for 13.9% of global emissions.

The main factors contributing to global warming are as follows: (1) human beings who bring about population explosion, air pollution, deterioration of the marine ecological environment, land salinization and desertification, sharp drops in forest resources, etc.; (2) nature, which has brought about volcanic activities, periodic fluctuations in the earth's orbital path, etc.

Approximately 75% of the respondents in the sampling survey expressed or implicitly held the view that most of the global warming observed over the past 50 years has been caused by human activities. However, some scientists assume that global temperature rise is still within the range of natural temperature changes; global temperature rise can be attributed to the arrival of the Little Ice Age; global temperature rise is also possibly due to changes in solar radiation and cloud cover adjustment; global temperature rise is an indicator of the urban heat island effect.

The United Nations Intergovernmental Panel on Climate Change predicts that human beings will encounter a warming world in the next half century or a decade. Due to the influence of human activities, in the twenty-first century, the concentration of greenhouse gases and sulfide aerosols has rapidly increased, leading to a rapid rise in global temperature in the coming century. As a result, the average global surface temperature will rise by 1.4–5.8 °C. The average temperature will rise by 2.2 °C by 2050 in China. Global warming is ringing an alarm to countries all over the world that climate change has seriously affected the survival of human beings and the sustainable development of society. Global warming is not only a scientific issue but also a comprehensive issue involving politics, economies, energy, and so on; thus, global warming is a national security issue.

Global warming has extremely serious consequences, which are listed as follows: (1) With global warming, the glaciers are melting, and the sea level is rising, leading to the loss of ecological communities such as coastal wetlands, mangroves and coral reefs, coastal erosion, saltwater intrusion into coastal, underground freshwater systems, coastal salinization, etc., which results in an imbalance of the natural environment in coastal areas, estuaries and bays; global warming brings great danger to the ecological environment in coastal areas. (2) Water areas are expanding. First, the larger the amount of water evaporated, the more the rainy season is prolonged, and floods are becoming more frequent. Then, the odds of flooding increase, and storms bring about more disasters. (3) With increased temperatures, the snow and ice in the Antarctic Peninsula and the Arctic Ocean are melting, contributing to the extinction of polar bears and walruses. (4) Many small islands will be submerged. (5) Original ecosystems are altered, having negative impacts on agriculture, forestry, animal husbandry, fisheries, etc.

6.1.3   Destruction of the Ozone Layer

The ozone layer is a region of the earth's stratosphere with a higher concentration of ozone, and the main function of the ozone layer is to absorb ultraviolet radiation. Ozone is created when ultraviolet light strikes ordinary oxygen molecules containing two oxygen atoms and splits them into individual oxygen atoms and each atom is then combined with unbroken oxygen. Most of the ozone layer is approximately 20–50 km above the earth. The ozone in the ozone layer is mainly formed by ultraviolet light reaction. On November 1, 2011, the Japan Meteorological Agency announced that the maximum area of the ozone layer over Antarctica detected by the agency exceeded that of 2010 and was equivalent to the average of the past decade.

At present, ozone layer depletion is one of the global environmental problems, and it has received more attention from countries all over the world since the 1970s. Since 1976, the United Nations Environment Programme (UNEP) has convened various international conferences and adopted a series of resolutions on ozone layer protection. Since the discovery of the depletion of ozone above Antarctica in 1985, the so-called Antarctic ozone hole, the international community has paid more attention to protecting the ozone layer.

6.1.4 Ocean Acidification

In 2003, the term "ocean acidification" first appeared in *Nature*, the famous British scientific journal. In 2005, James Hughes, an expert in disasters and emergencies, outlined the potential threats of ocean acidification: a biological extinction occurred in the oceans approximately 55 million years ago, and the culprit was CO_2 dissolved in sea water with an estimated total of 4500 billion tons; it took at least 100,000 years for the oceans to recover. On August 13, 2009, approximately 150 of the world's foremost marine researchers signed the *Monaco Declaration* in Monaco. The declaration reflects the close attention of global scientists to the serious damage to the global marine ecosystem from ocean acidification. In addition, this declaration notes that the rapid changes in seawater pH levels are approximately 100 times faster than the past natural changes. However, in recent decades, the rapid changes in marine chemicals have seriously influenced the marine organics, food webs, biodiversity, fisheries, etc. Additionally, this declaration calls for policymakers to make policies that keep CO_2 emissions within safe limits to avoid dangerous climate change, ocean acidification, and other problems. If CO_2 emissions in the atmosphere keep growing, by 2050, coral reefs will not survive in most waters, leading to permanent changes for commercial fishery resources and severely threatening the food security of millions of people.

In fact, ocean acidification involves sea water absorbing excess CO_2 from the air, resulting in a decreased pH. The power of hydrogen is generally expressed by pH, ranging from 0 to 14, which means that the acidity is the strongest when pH is 0 and the alkalinity is strongest when pH is 14. The pH of distilled water is 7, which means it is neutral. Sea water should be weakly alkaline, and the pH of ocean surface water is approximately 8.2. However, oceans acidify if excess CO_2 in the air is absorbed and dissolved into the ocean. Research indicates that excess CO_2 emissions due to human activities reduced the sea surface pH by 0.1 as of 2012, indicating that seawater acidity has increased by 30%.

In 1956, Logan Rowell, an American geochemist, started researching the effect of CO_2 produced during the industrial era on climate change in the following half century. Logan Rowell and his partners set up two monitoring stations in remote areas away from CO_2 emission points. One of the stations is in Antarctica, an area free of dust and industrial activities and deserted with almost no vegetation coverage; the other is at the top of Mauna Loa in Hawaii. Over half a century, they have monitored changes in CO_2 emissions and their impacts on climate change. According to the monitoring statistics, the annual CO_2 concentration has always been higher than the previous year, while the CO_2 emitted to the atmosphere has not been completely absorbed by plants and oceans. In fact, a considerable amount of

CO_2 remains in the atmosphere, and the amount of CO_2 absorbed by the oceans is very large.

Furthermore, the oceans and atmosphere are constantly exchanging gas, and any component that is released into the atmosphere is eventually dissolved into the ocean. Before the industrial era, the change in atmospheric carbon was mainly caused by natural factors, which then cause natural fluctuations in the global climate. Nonetheless, during the industrial era, human beings began to use a large amount of fossil fuels, including coal, oil, natural gas, and so on, but they also logged large stretches of forest. Therefore, by the beginning of the twenty-first century, humans had discharged more than 500 billion tons of CO_2, which makes the atmospheric carbon content increase year after year.

Under the influence of sea breezes, atmospheric components were absorbed into the sea surface for hundreds of meters deep; however, in the following centuries, these components have been gradually diffused to all corners of the sea bed. Research shows that during the nineteenth and twentieth centuries the ocean absorbed 30% of the CO_2 emitted by human beings, and now, it still absorbs these emissions at a rate of nearly 1 million tons/h. In 2012, US and European scientists found in their research that the ocean is experiencing the fastest acidification of the past 300 million years and that the acidification speed alarmingly exceeds that of the extinction 55 million years ago. There is no doubt that human activities have been acidifying the oceans. By 2100, the sea surface acidity is expected to drop to 7.8, and the seawater acidity will be 150% higher than that in 1800.

6.2 Influence of Global Environmental Change on Fisheries

6.2.1 Influence of Eutrophication on Fisheries

The damage from water eutrophication mainly manifests in the following three aspects: (1) Eutrophication causes decreased water transparency, which makes it difficult for sunlight to penetrate through the aqueous layer, which interferes with photosynthesis and oxygen release to aquatic plants. Meanwhile, plankton blooms consume much oxygen in the water, resulting in a serious shortage of dissolved oxygen, while photosynthesis of surface plants may cause supersaturation of partially dissolved oxygen. Either supersaturation or saturation decreasing the dissolved oxygen causes fish mortality. (2) Second, the harmful substances decomposed from the organics accumulated at the bottom of eutrophic water bodies under anaerobic conditions and the biotoxins released by some plankton are also harmful to aquatic animals. (3) Eutrophic water is rich in nitrite and nitrate poisons and is lethal for anyone who drinks it for a long period.

According to statistics, in 2012, 73 red tides were recorded in China's inshore waters, involving an area of 7971 km^2. The East China Sea witnessed the most frequent red tides among all affected areas, 38 red tides in total. The Bohai Sea witnessed the most extensive area affected by red tides among all affected areas,

3869 km^2 in total size. The most likely season for red tide occurrence is from May to June. There are 18 dominant species that induce red tides, which are *Karenia mikimotoi* Hansen, *Skeletonema costatum*, *Prorocentrum donghaiense*, *Aureococcus anophagefferens*, etc. From May 18 to June 8, 2012, in the coastal waters of Fujian Province in China, 10 red tides were documented as induced by *Karenia mikimotoi* Hansen, with a total affected area of 323 km^2. *Karenia mikimotoi* Hansen is an important toxic and harmful algae species contributive to red tides, as well as the main cause of the massive mortality of aquaculture shellfish (especially abalone) in Fujian Province in 2012. Red tides occur frequently in the inshore areas of Zhejiang Province, Liaoning Province, Guangdong Province, Hebei Province, and Fujian Province, while red tides occur most frequently in the inshore waters of central Zhejiang Province, Liaodong Bay, Bohai Bay, Hangzhou Bay, Pearl River Estuary, Xiamen, and northern Yellow Sea. Statistically, harmful red tides bring billions of dollars of economic losses to China's marine fisheries every year. Therefore, increasingly serious marine environmental pollution has destroyed the ecological environment of coastal and inshore fisheries to different degrees, causing a recession in traditional fishery resources, a shift in fishing grounds, and the disappearance of fish spawning grounds in estuarine and coastal waters.

6.2.2 Influence of Global Warming on Fisheries

6.2.2.1 Ecological and Physical Influences of Climate Change (Fig. 6.1)

With an increased global temperature, the amount of water vapor that evaporates from the ocean has greatly increased, which has aggravated ocean warming, while the geographical distribution of ocean warming is uneven. Climate warming causes the combined influence of temperature and salinity to reduce water density on the ocean surface, thus increasing vertical stratification. These changes may reduce the availability of surface nutrients, interfering with primary and secondary productivity in warm areas. Seasonal upwelling may be influenced by climate change, thereby affecting the whole food web. Climate warming results in changes in the community composition, productivity, and seasonal processes of plankton and fish. With ocean warming, chances are that more fish stocks will migrate to polar waters while less to equatorial waters. In general, climate warming is expected to drive the transformation of the distribution of most fish species to polar waters, the amplification of habitats for warm water species, as well as the shrinkage of habitats for cold water species. These changes also apply to pelagic species, which are expected to move to deeper waters to offset the temperature increase. Moreover, ocean warming will also change the predator–prey matching relation and then further change the entire marine ecosystem (FAO 2009).

Investigations have shown that global warming has caused the collapse of two ice shelves in the Antarctic, which gave rise to a seabed of $10,000$ km^2. As such, scientists have been able to discover many unknown new species of octopus,

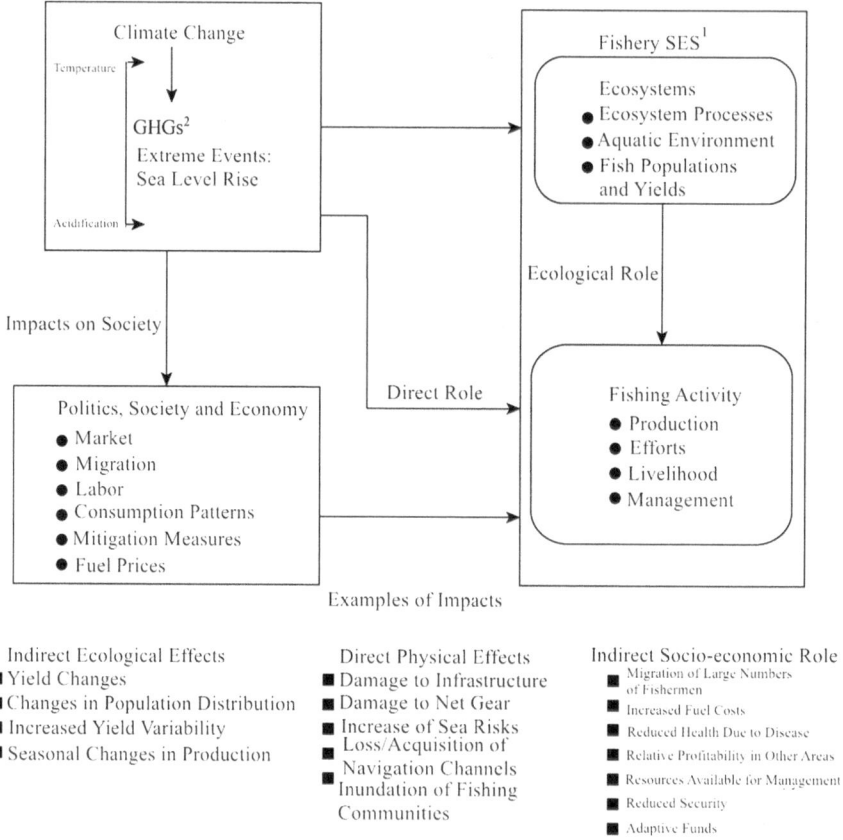

Fig. 6.1 Schematic diagram of the direct and indirect influences of climate change on fisheries (FAO 2009). [1]Social-ecological system; [2]Greenhouse gases

coral, shrimp, etc. The National Oceanic and Atmospheric Administration found that, over the past decade, stranding deaths of the Humboldt squid on the west coast of the USA have increased in number. In fact, the Humboldt squid generally lives in the warm waters south of the Gulf of California and along the coast of Peru. Nonetheless, this species swims northward and a large number die on the beach as the ocean becomes warmer. The northern range distribution of the Humboldt squid has also extended from 40°N in the 1980s to 60°N.

There is evidence that inland waters are also warming, and climate change has different influences on river run-off that flows into these waters. Generally speaking, lakes at a high latitude and high altitude are more likely to encounter a descending ice cap, warmer water temperature, and longer growing season; hence, the abundance and productivity of algae in these lakes will increase. On the contrary, probably because of a reduced nutrient supply, some deep lakes in tropical areas may experience algal blooms and reduced productivity. As for the average

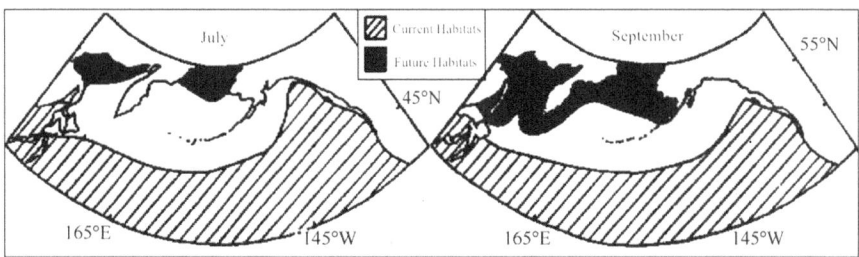

Fig. 6.2 Schematic diagram of the declining habitats for sockeye salmon (*Oncorhynchus nerka*) with a doubled CO_2 concentration under ocean warming (Williams 1988)

freshwater systems, considering the consequences of climate change, special attention should be paid to changes in the timing, intensity, and duration of flooding, as many fish species have to be adapted to the above changes as far as their migration, spawning, and spawning material transportation are concerned.

The United Nations Intergovernmental Panel on Climate Change found that the average temperature of the earth rose by 0.5–1 °C in the twentieth century because of the greenhouse effect, and great changes have occurred in the structure and function of the earth's ecosystems (including the fisheries ecosystem). In 50 or 100 years, the influence of climate change on world fisheries may be even more severe than that of overfishing. Fish are poikilothermic, so they have to migrate to new habitats, such as waters in higher latitudes or open waters, to survive when the former habitat becomes warmer under climate change. Global warming has few effects on fish habituated to the waters in middle and low latitudes, and the reasons are listed as follows: (1) Temperatures at the middle and low latitudes have comparatively changed within a small range during global warming. (2) The limiting factors of middle and low latitude fisheries production mainly include prey, red tides, and disease. In comparison, production of fisheries at high latitudes where light and temperature are the main limiting factors is more affected, while it is related to changes of physical factors (e.g., temperature, wind, current, salinity, etc.) of high latitude waters during global warming (Beamish 1993, 1995). Scientists from Canada, Japan, the UK, and the USA have conducted research into the dynamic relationship between the temperature of the cold temperate zone in the northern hemisphere and the habitats of sockeye salmon (*Oncorhynchusnerka*) in the last 40 years of the twentieth century (Williams 1988). With these research findings, the warming trend of the future sea surface temperature is predicted to drive sockeye salmon to migrate away from most of the North Pacific waters. If the sea surface temperature rises by 1–2 °C in the middle of the twenty-first century, their habitat will be confined to only the Bering Sea (Fig. 6.2). If these habitats are shrinking, the breeding migratory distance of the fish will be greatly extended, resulting in a smaller size of spawning brood stock and a decrease in spawning.

Therefore, the increased water temperature changes the spatial and temporal distribution of fish and their geographical populations. Meanwhile, this increase also brings long-term trend changes to the spatial and temporal distribution and

geographical composition of phytoplankton and zooplankton, aquatic primary producers, which ultimately has a lasting and profound impact on fisheries by bringing
about structural changes of the upper food webs that consume phytoplankton as
prey, thus having a profound influence on fisheries.

6.2.2.2 Changes to Fisheries in Different Marine Ecosystems

Global climate change (such as El Niño-Southern Oscillation (ENSO) and extreme
weather events) affects the abundance and distribution of fishery resources and the
suitability of geographic locations of aquaculture farms for cultured species. Factors
influencing the global ocean (including coastal areas) and freshwater systems
include chemical factors (including salinity, oxygen content, carbon sequestration,
acidification, etc.) and physical factors (including temperature, water level, ocean
circulation, wind system, etc.). Factors influencing organisms in different water
systems also vary. For instance, the main influencing factors for a marine system
could be temperature, sea level, circulation, waves, salinity, oxygen content, carbon,
and acidification, while the main influencing factors for freshwater systems could be
evaporation and precipitation, temperature, and storms. In addition, fish and shellfish
react differently in different ecosystems under climate change. That is to say, global
sea surface warming, anoxic zone diffusion, and pH decline can change biological
systems by changing the abundance composition, individual size, trophic relationships, and interaction dynamics.

Global marine ecosystems can be categorized as high latitude algal bloom system,
coastal boundary systems, western boundary ascending system, equatorial upwelling
systems, semienclosed seas, and subtropical circulations (Fig. 6.3; Guldberg et al.
2014). The ocean is a major contributor to global food security, as more than 80% of
marine catches are from the northern high-latitude spring bloom system, the coastal
boundary system, and the western boundary upwelling system. Marine species and
ecosystems are affected by climate change, including the migration of marine
organisms to higher latitudes, the faster movement of fish and zooplankton in
high-latitude systems, etc.(Guldberg et al. 2014). Moreover, climate change influences fish stock abundance, mainly by affecting supplementary populations. Therefore, the influencing factors and mechanisms for fish vary in different marine
ecosystems. In short, the influence of climate change on fish mainly includes:
(1) direct influences that change the metabolism or reproduction process in fish
and; (2) indirect influences that change the biological environment of fish, including
related prey, predators, species interactions, diseases, etc.

6.2.2.3 Influence of Climate Change on Fishermen and their Communities

The economy, coastal communities, and fishermen dependent on fisheries are
expected to be influenced by climate change in different ways. For example, people

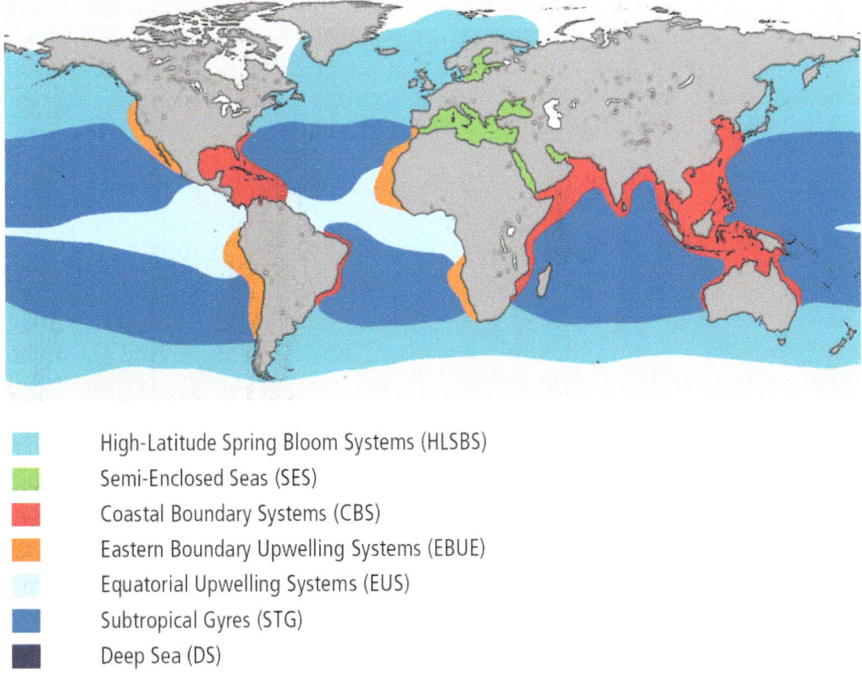

High-Latitude Spring Bloom Systems (HLSBS)
Semi-Enclosed Seas (SES)
Coastal Boundary Systems (CBS)
Eastern Boundary Upwelling Systems (EBUE)
Equatorial Upwelling Systems (EUS)
Subtropical Gyres (STG)
Deep Sea (DS)

Fig. 6.3 Schematic diagram of marine ecosystem distribution (Guldberg et al. 2014)

migrate due to sea level rise and tropical storm frequency, distribution, or intensity, which would influence coastal communities and infrastructure; livelihoods become unstable, and changes take place as to the availability and quantity of fish as food.

First of all, the vulnerability of fisheries and fishing communities is determined by the extent of their exposure to changes, their sensitivity to changes, and the capability of an individual or system to predict. The capacity to adapt is determined by community assets, which are either influenced by culture, current institutional and governance frameworks, or by exclusion from the use of adaptive resources. Vulnerabilities vary among countries and communities and within communities. Generally speaking, poorer and less powerful countries and individuals are more vulnerable to climate change, and the more vulnerable are fishing communities where resources are affected by overfishing and ecosystem degradation and where social services and essential infrastructure are lacking.

Additionally, fisheries are a dynamic social–ecological system that goes through rapid changes in markets, development, and governance. Whereas, the combined effects of these changes and climate change on nature make it difficult to predict the future influence of climate change on fisheries social–ecological systems.

In contrast, human adaptation to climate change involves responses or projected actions by individuals or public institutions. For example, fishermen can give up fishing that has been an alternative job for them; safety guarantee and disaster forecasting systems can be established; and the type of fishing practice can be adjusted. As far as fisheries governance is concerned, changes in stock distribution and abundance index have to be handled, thus establishing rational and sustainable fisheries. Acceptance of uncertainties and an ecosystem-based approach are the best way to improve the adaptive capacity of fisheries.

6.2.3   Influence of Ozone Layer Destruction on Fisheries

Once the ozone layer is depleted, its ability to absorb ultraviolet radiation will be greatly reduced, which will lead to a significant increase in ultraviolet light B that reaches the earth's surface. Then, this increase brings harm in many aspects to human health and the ecological environment, including terrestrial plants, aquatic ecosystems, biochemical cycles, materials, troposphere atmospheric composition, air quality, and so on. The problem of ozone layer depletion has received much attention worldwide.

Although the ozone layer depletion crisis is not as obvious as environmental pollution, its impact on life is very significant. If the ozone layer is depleted, the ultraviolet rays from the sun can easily invade the earth, causing a catastrophe for natural ecology and human beings. For example, ultraviolet radiation is so strong that it causes lethal damage to plant growth, which in turn, affects the ecology of the earth; meanwhile, excessive ultraviolet radiation kills plankton on the ocean surface, and once these organisms at the bottom of the food chain die, then the balance of the entire marine ecosystem is damaged.

In conclusion, ultraviolet light will decrease crop production and cause food shortages. According to scientific observations, the amount of ultraviolet radiation increases when the concentration of the ozone layer decreases. For example, when the concentration of the ozone layer decreases by 1%, the yield of soybeans will decrease by 1%, and the soybean quality will also be relatively poor. As a result, ultraviolet radiation destroys plant-based production on land and in the ocean, leading to the death of animals due to a lack of food, leading to ecosystem imbalance.

Researchers have measured the amount of ultraviolet-B (UV-B) radiation in Antarctica and the increased amount penetrating into the water, so there is sufficient proof that natural phytoplankton communities are directly related to ozone changes. Based on a comparison of phytoplankton productivity within and outside the ozone hole, scientists have found that the decline in phytoplankton productivity is directly related to increased UV-B radiation caused by ozone reduction. In addition, a study has shown that productivity at the glacier marginal areas has decreased by 6%–12%. Plankton plays an important role in the marine food chain, so a reduction in the species and population of plankton also influences the production of fish and shellfish. Another scientific study has shown that if stratospheric ozone decreases by 25%, the primary productivity of plankton will decline by 10%, which will result in a 35% reduction in the population of organisms living near the water surface.

According to the study, UV-B radiation in sunlight has harmful effects on the early development of fish, shrimp, crabs, amphibians, and so on, among which the most serious is decreased fertility and larval immaturity. UV-B has already been a limiting factor for life on earth. Moreover, a slight increase in UV-B exposure results in a significant reduction in the number of consumer organisms.

In a long run, the number of biotic populations that are sensitive to ultraviolet radiation is inevitably inhibited, and the interspecific competitiveness of organisms that are insensitive or rehabilitative will be strengthened, eventually resulting in structural changes in aquatic ecological communities. To date, the impact of this change on fisheries production is not yet known, but from a long-term perspective, it is possible to entirely exceed the direct inhibition of UV radiation on basic productivity.

6.2.4 Ocean Acidification and Its Impact on Fisheries

6.2.4.1 Impacts on Phytoplankton

Phytoplankton constitutes the foundation and primary productivity of the marine food web, so their "reshuffle" is likely to negatively impact many marine animals ranging from small fish and shrimp to large sharks and whales. Moreover, the prey value of nutrients will decrease as the seawater pH becomes lower and will also change the capability of phytoplankton to absorb various nutrients, and the increasingly acidified sea water will corrode the bodies of marine organisms. According to research, in an acidified environment, the efficiency of calcified organisms, such as algae, anthozoans, shellfish, crustaceans, and echinoderms, to form calcium carbonate shells and skeletons has been significantly reduced. Due to global warming, the density of the ocean surface layer that absorbs atmospheric CO_2 also decreases with the temperature rise, thereby weakening the material exchange between the surface layer and the middle and deep sea water and attenuating the upper mixed layer of the ocean, which is not conducive to phytoplankton growth.

6.2.4.2 Impacts on Coral Reefs

Research shows that tropical coral reefs provide shelter, food, and breeding sites for nearly 25% of fish, which account for 12% of the global catch. Additionally, coral grows best when the average seawater pH is 8.1; *Alcyonium* becomes the best growing coral when the seawater pH is 7.8; if the pH drops below 7.6, neither of them could survive. The pH of natural sea water is stable, ranging between 7.9 and 8.4, while the pH of unpolluted sea water is between 8.0 and 8.3. In fact, the weak alkalinity of sea water is beneficial for marine organisms that use calcium carbonate to form shells. Japanese scientists have found that the seawater pH is expected to reach 7.8 at the end of the twenty-first century, the acidity will rise sharply compared with the normal state, and corals may disappear.

6.2.4.3 Impacts on Mollusks

Some research suggests that by 2030, the oceans in the southern hemisphere will have a corrosive effect on snail shells, which are important food sources for salmon in the Pacific Ocean, so if snails are depleted in some waters, it will influence the salmon fisheries. In addition, the inner shell of squid will thicken and increase in density in an acidified ocean, which will slow the squid swimming speed and affect its ingestion and growth.

6.2.4.4 Impacts on Fish

Based on experiments, scientists have found that if the same batch of fish is left in normally acidified sea water, only 10% of them are captured in 30 h; however, if they are left in the acidified waters near the Great Barrier Reef, with other conditions remaining the same, they are all captured by nearby predators within 30 h. According to the latest reports from the *Proceedings of the National Academy of Sciences of the United States of America*: in acidified waters simulated as the waters in the next 50–100 years, fish instinctively avoid predators in the sea water with highest acidity, but they are then attracted by the smell of the predator because their olfactory system has been destroyed.

6.2.4.5 Impacts on Marine Fisheries

First of all, ocean acidification has direct effects on the quantity and quality of marine living resources, leading to permanent changes in commercial fishery resources, which ultimately affects the production and output value of the marine fishing industry, as well as threatens the food security of millions of people. Although there are no convincing predictions about how much changes in the chemical properties of sea water will affect fisheries production, ocean acidification will definitely result in a decline in fisheries production and an increase in fisheries production costs.

Second, ocean acidification leads to reduced fish habitats. In the Pacific region, coral reefs are the main habitat for fish and other marine animals that provide approximately 90% of the protein for Pacific island countries. It is estimated that coral and coral ecosystems create more than $375 billion in value for human beings every year. Therefore, if the coral reefs are substantially reduced, it will have a significant impact on the environment and social economy.

Third, ocean acidification leads to decreased fish food. Ocean acidification can hinder the ability of certain plankton to form calcium carbonate, while plankton is at the bottom of the food chain and large in quantity; ocean acidification hinders the growth of these organisms, resulting in reduced production of fish, which are in the upper layers of the food chain.

The FAO estimates that more than 500 million people worldwide rely on fishing and aquaculture for protein intake and as an income source. For these people, fish provide approximately half of the amount of animal protein and trace elements needed daily by the poorest 400 million people. In conclusion, the impact of ocean acidification on marine life will inevitably endanger the livelihoods of these poor people.

6.3 Impact of Global Climate Change on Aquaculture

6.3.1 Overview of the Impact of Global Climate Change on Aquaculture

Currently, aquaculture products account for nearly half of human consumption of aquatic products, which is expected to increase further to satisfy additional demand. Aquaculture is mainly distributed in the tropical and subtropical regions of the world, and inland freshwater production in Asia accounts for 65% of total production. Therefore, many aquaculture activities are carried out in the deltas of main rivers. Moreover, climate change will have a series of influences on aquaculture; thus, it is necessary to understand the driving factors, influence pathways, variability, and risks of climate change when policy makers create the adaptation strategies for this industry.

Above all, research has defined the main factors that may have a direct or indirect influence on aquaculture, and this research has described the empirical evidence for these influences. The influence factors include water warming, sea level rise, ocean acidification, changes in climate patterns, and extreme weather events. The Fifth Assessment Report of the IPCC provides empirical evidence for the influence of global warming on oceans and coastal and inland water bodies. Therefore, coastal systems and lower lying areas will be affected by the increasingly severe risks of flooding, seawater intrusion, coastal erosion, saltwater intrusion, etc. As a result, coastal systems are confronted with the most serious risks.

In several studies, the correlation between each influence factor and its impact on aquaculture has been roughly determined, some factors have been confirmed and the correlations vary in intensity. For instance, the prediction of an elevated CO_2 concentration in sea water and the resulting acidification problem will physiologically influence the growth and reproduction of bivalves and possibly affect shell quality. However, climate warming will also increase the spatfall and growth rate of shellfish, thereby expanding the latitude of aquaculture farms, so climate change may also bring about benefits. On the contrary, aquaculture practitioners and researchers have reported that a large number of oyster seedlings in hatcheries have died from elevated water acidity. Therefore, more studies should be conducted to clarify the influences of acidification on marine finfish. Embryos and larvae appear to be more sensitive to elevated CO_2 concentrations than juvenile and adult fish; thus, the concentrations may have sublethal effects, such as a slower growth rate. There is a correlation between climate-related changes (e.g., temperature and growth rates, disease susceptibility, spawning time, and mortality at specific stages of the life cycle) and the direct economic influence in the culture process. Lastly, extreme weather events will impact metabolic reactions through changes in salinity and temperature, thus leading to physiological effects and longer-term physiological changes. In addition, these changes will also bring about various socioeconomic influences, including escape from aquaculture farms, destruction of infrastructure, other livelihood assets, and so on.

Then, climate change will also have an indirect impact on aquaculture by influencing feed, seedlings, fresh water, and other inputs, which includes fishmeal fisheries, wild seedlings, and land-based feed such as soybeans, maize, rice, and wheat. Moreover, climate change will have an indirect impact on fish disease. The Fifth Assessment Report recognized that the threat of fish disease to aquaculture is increasing with global climate changes, and many researchers have studied the spread and outbreak of disease on aquaculture organisms as well as the distribution of parasites and pathogens. For example, Vibrio may be a disease that is severely affected by climate change because *Vibrio* prefers to grow in warm (>15 °C) and low salinity (<25) water. In temperate and frigid zones, the *Vibrio* eruption in soft shellfish has been associated with climate warming. Because the aquaculture environment of fish and shellfish, especially ponds or circulatory systems, can be adjusted within a certain range, it seems quite possible to cope with climate-related risks with artificially regulated environments, though additional costs are required. However, the global aquaculture industry is dominated by small- and medium-sized farmers, and they have relatively limited abilities to regulate farming systems.

6.3.2   Impact of Climate Change on the Vulnerability of Regions, Countries, and Aquaculture Species

6.3.2.1   Impact on the Vulnerability of Aquaculture in Different Regions and Countries

According to the predictions provided by the *Fifth Assessment Report* issued by IPCC, tropical ecosystems are highly vulnerable to climate change; thus, climate change negatively impacts communities that depend on tropical ecosystems. Climate change will impact food security in Asia by the mid-twenty-first century; South Asia will be the region suffering the most serious food security problem. In fact, nearly 90% of the world's aquaculture is conducted in Asia, mostly in the tropical and subtropical regions. A study, using a series of indicators in a GIS model (such as exposure, sensitivity, and adaptability), has identified Bangladesh, Cambodia,

China, India, the Philippines, and Vietnam as the most vulnerable countries in the world. Recently, another study has assessed this vulnerability again using more advanced models and data and concluded that most aquaculture countries in Asia are very vulnerable. However, after considering all the water environments (fresh water, brackish water, and salt water), Bangladesh, China, Thailand, and Vietnam are identified as the most vulnerable countries. However, in regions outside Asia, Costa Rica, Honduras, and Uganda are among the 20 most vulnerable countries as far as freshwater aquaculture is concerned; as for brackish aquaculture, Ecuador and Egypt are very vulnerable; as for marine aquaculture, Chile and Norway are vulnerable. Among all these vulnerability models, sensitivity is estimated by the contribution of aquaculture production and aquaculture to GDP (gross domestic product), but for those countries where aquaculture has only started and has great development potential, such as African countries, researchers ignore sensitivity, so a relative vulnerability estimate is put forward.

6.3.3 Impact on the Vulnerability of Species and Systems

Several different approaches can be used to assess the vulnerability of individual species and systems when institutional and structural adaptation strategies are designed at the fishermen and local levels. However, perhaps, the most practical approach is classifying aquaculture activities by geographic factors, including inland, coastal, or arid tropical areas, etc., and then further classifying them by farm density and production intensity. For the same cultured species at the same site, the factors affecting the system vulnerability include technology, aquaculture management measures, and regional management.

Compared to large commercialization stakeholders, poor and small-scale stakeholders are relatively disadvantaged in grasping opportunities and responding to threats. Therefore, more attention should be paid to developing the overall adaptability, supporting the poor and small-scale aquaculture farmers, and value chains; each party along a value chain should take maximum advantage of new opportunities to address the challenges of climate change.

6.3.4 International Actions to Reduce Climate Change Impacts on Aquaculture

Some practical adaptation measures can effectively address the problems brought about by climate variability and climate trends to aquaculture farms at the local, national, and even global levels. Aquaculture farmers and other local stakeholders can play an active role in addressing long-term changes/trends and sudden changes (such as extreme weather events) with these measures, and their role is as follows:

developing aquaculture zoning to minimize risks by relocating to less risky areas; carrying out appropriate fish health management; enhancing water use efficiency, water resource recycling, aquaponics, etc.; improving feeding efficiency to reduce pressure and dependence on feed resources; developing more adaptable species (including species with low pH and salt tolerance but fast growth, etc.); ensuring that hatching production is high quality, reliable, and easy to grow under tough conditions, promoting postdisaster recovery; improving monitoring and early warning systems; reinforcing farming systems, including improved farming facilities (such as stronger cages, depth-adjustable cages to adapt to water level fluctuation and deeper culture ponds) and management measures; and improving fishing methods and value-added activities.

In fact, some countries have started taking action. For example, Vietnam has taken actions to breed squid species with better salt tolerance, and the Bangladesh government and its partners are exploring various options, including culturing salt-tolerant species, deepening ponds, using depth-adjustable cages, applying polyculture of fish and crops, etc.

Then, to monitor the implementation of the 1995 *Code of Conduct for Responsible Fisheries*, the FAO issued a questionnaire to member countries specifically concerning the aquaculture industry. The questionnaire was designed to assess the measures taken by an institution to adapt to climate change and the resilience governance for climate change (Table 6.1; FAO 2016). This latest assessment highlights multiple weaknesses in institutional and governance responses to climate change, especially in regions where the aquaculture industry is still in its infancy. To be better prepared to mitigate climate change risks, governments must first fully understand the vulnerability of aquaculture at the local and national levels. However, this is still a global gap. Additionally, it should be seen as a priority to strengthen preparedness and promote the development of adaptation measures.

Aquaculture zoning, one of the key measures, is very weak on a global scale, especially in regions where the aquaculture industry is still in its infancy. In fact, the conditions of aquaculture facilities directly determine the risk exposure degree, thereby also determining vulnerability. For instance, when selecting the location of fish cages in coastal areas, the following factors need to be considered: the degree of the effect of weather events; changes in water flow or sudden influxes of upstream fresh water; long-term trends such as rising temperature, rising salinity, and decreased oxygen level.

The above information is of vital importance to determine aquaculture zoning and farm location. Nevertheless, as for the spatial distribution of inland and coastal aquaculture ponds, people consider whether it is more convenient to obtain more land and water resources in many parts of the world instead of avoiding the impact of external threats. Integrating climate change and other risks into spatial planning and aquaculture zoning is an urgent task for regions and countries where the aquaculture industry is still in its infancy. However, for the areas where it is difficult to relocate aquaculture systems, it is very necessary to take into account the concept of risk-based regional management. Government disaster assistance and farmer access to

Table 6.1 Average scores in the 2015 code questionnaire on aquaculture on the presence of measures for reducing vulnerability to climate change (FAO 2016)

Project	Africa	Asia	Europe	Latin America and the Caribbean Region	Near East	North America	Southwest Pacific	Global
Number of countries	14	10	18	19	5	2	2	70
Overall preparedness for managing climate change-related risks	1.7	2.7	2.9	1.6	2.6	3.5	3.0	2.3
Overall preparedness for disaster response	2.2	2.9	3.1	2.2	2.6	4.0	3.0	2.6
Aquaculture zoning for dealing with production, environmental and social risks	2.6	3.0	2.6	2.4	3.0	3.5	4.0	2.5
Assistance provided by the government for farms under disaster conditions	2.3	1.9	1.1	1.3	2.0	0.0	1.5	1.2
Commercial insurance offered to farmers	1.3	1.3	1.1	1.3	0.3	0.0	1.0	0.8
Implemented fish hygiene management	2.7	3.5	4.0	3.2	3.2	4.5	3.5	3.3
Institutional credit and microcredit obtained by farmers	2.8	1.3	1.2	1.5	2.5	0.0	1.0	1.2
Integration of aquaculture into coastal management plans	2.8	3.7	2.9	2.5	2.6	3.5	3.5	2.6
Integration of aquaculture in watershed management or land use development plans	2.4	3.3	2.9	2.1	3.6	3.5	2.0	2.5
Consideration of ecosystem function in aquaculture planning and development	2.4	3.8	3.6	2.6	2.4	4.0	3.0	2.9
Established incentive mechanisms to encourage farmers to restore ecosystem services and resources	1.8	2.7	1.7	1.8	2.0	4.0	3.0	1.5
Implemented best management practices (BMPs)	2.5	4.0	3.0	3.0	2.8	4.5	3.0	3.0

Note: Score for each item is between 0 (there is no such measure) and 5 (the measure has been established, fully implemented, and put in place throughout the country) commercial insurance, the two other measures, are extremely limited in Asia, one of the major aquaculture-producing areas with the most vulnerability.

Fish disease is one of the common reasons for significant losses in aquaculture, so adequate fish health management and biological safety are key to the resilience of this industry. The preparedness for fish disease response scored higher than other measures on a global scale, indicating that this measure was implemented well. However, climate change may increase the incidence of disease and its influence, so implementation must be further strengthened, especially in Asia, where aquaculture is more intensive and the density of farms per unit area is higher.

Although progress has been made in understanding the vulnerability of aquaculture under climate change, more research is needed to identify the driving processes and develop alternative aquaculture methods and measures. However, decision making and planning does not mean waiting for the progress of knowledge. On the contrary, the main challenges must be actively addressed with existing knowledge; adaptation strategies should be developed to minimize vulnerability under climate change. Many measures are one part of the best measures available in aquaculture. Therefore, as stakeholders do not make major changes in a general direction, they only focus on priority areas. For example, stakeholders must increase their focus on aquaculture zoning to better address climate change to ensure that farms are built in low-risk areas or to encourage farms in high-risk areas to take measures (e.g., building deeper breeding ponds, introducing more resilient species, etc.).

As for the local level, one of the practical adaptation measures is local environmental monitoring. Aquaculture is extremely sensitive to sudden climate change and long-term climate trends. However, except for some industrial aquaculture, currently, there has not been a comprehensive monitoring system that provides farmers with information that can be used in decision-making processes. In fact, long-term collection of simple data (such as fish behavior, salinity, water temperature, transparency, and water level) can provide useful information for decision making, especially when changes can have serious consequences. In addition, the information collected and shared locally helps farmers to better understand biophysical processes and participate in the process of finding solutions, including rapid adaptation measures, advanced warning, long-term behavior, investment changes, etc. To implement a monitoring system, the following activities are required: provide training to local stakeholders to learn the value of monitoring and how to take advantage of the monitoring results for decision making. Moreover, it is necessary to establish a simple network/platform for receiving, sharing, and analyzing information; coordinating and interconnecting with other forecasts; and providing feedback to local stakeholders in real time.

6.4   Food Security Vulnerability Assessment of Marine Fisheries under Climate Change

6.4.1   Overview of Research

Fisheries management used to be focused on maximizing the benefits of capture fisheries in all aspects of employment, income, and export, while simultaneously ensuring the sustainability of fishery resources. However, recently, people have transferred attention to using fish as an important food source and a necessary nutrient while ensuring ecosystem protection. The above transformation is manifested in the fact that the Sub-Committee on Aquaculture and the Sub-Committee on Fish Trade of the FAO Fisheries Commission have chosen fish and nutrition as topics for recent meetings. Vulnerability assessments under the effects of climate change on marine fisheries can help reduce vulnerability, because they identify countries that are most vulnerable to climate change, countries where food security, employment, and the economy are highly dependent on the fisheries sector and countries with low adaptive capacity and limited resources and social capacity. Overall, vulnerability assessment plays a core role in identifying areas where measures are most needed to prioritize the implementation of climate adaptation plans, and these assessments have now received increasing attention from policy makers and academia.

Many scholars have carried out fisheries vulnerability assessments on different scales since the first global vulnerability assessment of fisheries in 2009. In fact, vulnerability assessments carried out at the national scale can identify the most vulnerable countries and thus provide guidance for policy responses and adaptive management strategies on a national scale. According to statistics, in 2013, 57% of the world's total production of fish, crustaceans, and mollusks was from capture, and 43% of it was from aquaculture. In addition, among the total global catch in 2013, 87% of it was from marine waters and 13% of it from inland waters. The proportion of global marine catches to total production varies greatly on a national scale. Therefore, although the vulnerability to climate change in some countries is at the same level, the impact of food security varies in different countries. Ding and Chen (2017) have assessed the vulnerability of coastal countries to climate change using four environmental indicators that have a more direct and significant influence on marine fisheries: (1) sea surface temperature anomalies; (2) ultraviolet radiation (UV radiation); (3) ocean acidification; and (4) sea level rise.

6.4.2   Vulnerability Assessment of Global Fisheries

Ding and Chen (2017) have assessed the food security vulnerability of marine fisheries under climate change on a national scale. They found that the food security vulnerability of marine fisheries under climate change is closely related to the state of

Table 6.2 Coefficient of determination of the vulnerability of National food security under climate change as predicted by indicators (Ding and Chen 2017)

Index	Value of R^2
Exposure degree	0.42
Sea surface temperature anomaly	0.07
Ocean acidification	0.26
Ultraviolet radiation	0.09
Sea level rise	0.19
Sensitivity	0.57
Food dependence	0.31
Employment dependence	0.30
Economic dependence	0.26
Adaptability	0.64
GDP per capita	0.46
Life expectancy at birth	0.49
Global governance index	0.50
Human development index	0.64

national development. First, developing regions, such as Africa, Asia, Oceania, and South America, are the most vulnerable. Vulnerability is most significantly correlated with adaptation ($R^2 = 0.64$), followed by sensitivity ($R^2 = 0.57$), while vulnerability is less correlated with exposure ($R^2 = 0.42$) (Table 6.2). Among the 27 highly vulnerable countries, 23 are also highly sensitive countries. The independent variable most associated with vulnerability is the Human Development Index (HDI) ($R^2 = 0.64$), followed by the World Governance Index ($R^2 = 0.50$) and Life Expectancy at Birth ($R^2 = 0.49$) (Table 6.2). In conclusion, countries with higher levels of development, stronger governance, and a longer life expectancy will have lower food security risks caused by climate change.

First of all, the food security vulnerability of marine fisheries under climate change is the combined result of the exposure degree, sensitivity, and adaptability. European countries have lower exposure and sensitivity but higher adaptability; therefore, there are no European countries highly vulnerable to climate change. Furthermore, Iceland is highly sensitive to climate change, mainly because its marine fisheries have a high contribution to the national GDP. Bulgaria, Greece, Iceland, and Romania have moderate exposure degree, which are compensated by their lower fisheries dependence and higher adaptive capacity, making the vulnerability of these countries low or extremely low.

Second, North American countries have relatively low dependence on marine fisheries, so there are no high-sensitivity countries in North America. Barbados, the Dominican Republic, and Trinidad and Tobago have moderate adaptability (these countries have a relatively high life expectancy at birth, high global governance index, and high level of economic development), which partially offsets their high exposure degree. The extremely low adaptability and moderate exposure have resulted in the moderate vulnerability of Honduras. Moreover, there are seven North American countries that are also moderately vulnerable, but the implicit factors that contribute to the vulnerability of these countries are different. For instance, Cuba, Jamaica, and Saint Vincent and the Grenadines are moderately vulnerable, mainly because of their high exposure and heavy reliance on marine fisheries for employment opportunities, while Belize, Honduras, and Nicaragua have medium vulnerability, mainly due to their higher vulnerability caused by low levels of GDP per capita under a high exposure degree.

The high exposure degree, high sensitivity (mainly due to high employment dependence), and extremely low adaptability of Guyana in South America lead to its high vulnerability. Although Venezuela also has a high exposure degree, its low sensitivity and moderate adaptability place its climate change vulnerability at a moderate level. Additionally, the value of marine fisheries in Peru accounts for 11% of its total GDP, which leads to its high sensitivity.

Third, most African countries have a high level of vulnerability. According to the research of Ding and Chen (2017), five out of the nine highly vulnerable countries are in Africa. Among the 109 countries examined in their research, Mauritania and Mozambique are the most vulnerable. In fact, the high vulnerability of African countries is mainly due to their high degree of exposure, fisheries dependence, and low degree of adaptability. African countries heavily rely on marine fisheries for employment opportunities, income enhancement, and food supply. For instance, the number of people engaged in marine fisheries in Guinea-Bissau accounts for 59% of the total economically active population; Mauritania's marine fisheries account for 23% of its national GDP; and the animal protein intake per capita in the Democratic Republic of the Congo is only 4.3 g, of which 38% is animal protein from fish. In addition, among the 27 extremely low-adaptive countries, 20 are in Africa.

Fourth, many Asian countries are highly vulnerable, such as Bangladesh, Cambodia, the Maldives, the Philippines, Vietnam, Thailand, and Indonesia. Among the above countries, Bangladesh, Cambodia, the Maldives, the Philippines, and Vietnam are highly dependent on marine fisheries, which are believed to provide employment opportunities, generate income, and supply animal protein. Among the 25 Asian countries, 10 have a high exposure degree, while the high exposure degrees in Cyprus and Israel are partially offset by their low fisheries dependence and high adaptability.

Fifth, except for New Zealand, the exposure degree of other Oceania countries is relatively high. The high exposure degree of Australia is compensated by its extremely low sensitivity and high adaptability. However, Fiji, Samoa, the Solomon Islands, and Vanuatu have a high exposure degree, and these countries are highly dependent on marine fisheries for protein and livelihood sources, while they have lower adaptability.

6.4.3 Prospects in the Future

Good knowledge of the countries that suffer the most climate change impacts and the reasons for their climate change vulnerability is a good niche to conduct future research and decision making. The national (fisheries-related) food security vulnerability caused by climate change is closely related to national development and also to the finding that developing countries in Africa, Asia, Oceania, and South America are the most vulnerable to climate change. Countries where marine fisheries are more likely to be affected by climate change and that have the highest food security vulnerability are highly dependent on marine fisheries to meet the nutritional needs of their people. In these African countries (Cape Verde, the Gambia, Guinea, Guinea-Bissau, Mauritania and Senegal), Asian countries (the Maldives), Oceania countries (Fiji, Samoa, the Solomon Islands and Vanuatu), and South American countries (Guyana), domestic aquatic production is the main source of fish protein supply, and marine catches account for more than 85% of total production.

Adaptability plays an important role in reducing the risks of food security, so countries that have high adaptability are less affected by environmental fluctuations and better able to seize opportunities to ensure national food security. Moreover, the vulnerability index is closely related to the Human Development Index, World Governance Index, and Life Expectancy at Birth. Therefore, the above three aspects may benefit the promotion of food security policy the most. Specific measures include increasing investment in education, improving governance, reducing poverty, and improving the adaptive capacity and health of fishermen.

Policies to stabilize aquatic products trade, food production, and food security are often developed and implemented at the national level, so it is important to conduct vulnerability assessments at the national level. Ding and Chen (2017) conducted a systematic and creative assessment on the food security vulnerabilities of marine fisheries under climate change in 109 countries around the world. This research found that developing countries in Africa, Asia, Oceania, and South America are the most vulnerable, which results from different factors in these countries. In two-thirds of countries where climate change affects marine fisheries and food security is highly vulnerable, marine fisheries serve as their main source of aquatic products supply. Therefore, it is of vital importance to maintain national food security in countries with high vulnerability and high dependence on marine fisheries and to reduce climate change impacts by developing appropriate adaptation strategies and management measures. Additionally, it is also vitally important to conduct vulnerability assessments on marine fisheries under climate change and to have knowledge of countries highly vulnerable to climate change, which is believed to facilitate the adoption of appropriate management measures to mitigate the influence of climate change and promote national food security.

References

Beamish RJ (1993) Climate change and exceptional fish production off the west coast of North America. Can J Fish Aquat Sci 50:2270–2291
Beamish RJ (1995) Response of anadromous fish to climate change in the North Pacific. In: Peterson DL, Johnson DR (eds) Human ecology and climate change: people and resources in the far north. Taylor and Francis, Washington, DC, pp 123–136

Ding Q, Chen XJ (2017) Assessment of sustainable development of global marine fisheries resources on basis of catch statistics. China Science Press, Beijing. (in Chinese)

FAO (2009)Climate change implications for fisheries and aquaculture- Overview of current scientific knowledge.FAO Fisheries and Aquaculture Technical Paper 530. Rome

FAO (2016) The state of world fisheries and aquaculture 2016. Rome

Guldberg O, Cai R, Poloczanska ES et al (2014) The ocean. In: Climate change 2014: impacts, adaptation, andvulnerability. Part B: regional aspects. Contribution of working group II to the fifth assessment report of the intergovernmental panel on climate change. Cambridge University Press, Cambridge, United Kingdom and NewYork, NY, USA, pp 1655–1731

Williams N (1988) Temperature rise could squeeze salmon. Science 280:1349